D1030112

FRAMING
FLOORS · WALLS · CEILINGS

THE BEST OF
Fine Homebuilding

FRAMING
FLOORS · WALLS · CEILINGS

THE BEST OF
Fine Homebuilding

The Taunton Press

Cover photo: Charles Miller

**Back-cover photos: Charles Miller (top and center),
Paul Spring (bottom)**

Taunton
BOOKS & VIDEOS
for fellow enthusiasts

First printing: 1996
Printed in the United States of America

A Fine Homebuilding Book

Fine Homebuilding® is a trademark of The Taunton Press, Inc.,
registered in the U.S. Patent and Trademark Office.

The Taunton Press, Inc.
63 South Main Street
P.O. Box 5506
Newtown, Connecticut 06470-5506

Library of Congress Cataloging-in-Publication Data

Framing floors-walls-ceilings : the best of Fine homebuilding.
　　　　p.　cm.
　"A Fine homebuilding book."
　Includes index.
　ISBN 1-56158-148-8
　1. Framing (Building).　　I. Taunton Press.　　II. Fine homebuilding.
TH2301.F74　　1996
694'.2 — dc20　　　　　　　　　　　　　　　　　　96-10779
　　　　　　　　　　　　　　　　　　　　　　　　　　　　CIP

CONTENTS

INTRODUCTION

I LEARNED TO FRAME HOUSES from a guy named Ed who had been a tank commander in the army. He used a 28-oz. framing hammer—anything less was a tack hammer—drove a dump truck instead of a pickup and prided himself on his ability to say "hurry up" in five different languages. Whenever I got in his way, which was fairly often, he referred to me as "a nickel holding up a dime."

I can remember framing with Ed on summer days that were so hot that we started at 6:00 a.m. just to beat the heat. And I remember the glare of the sun on the plywood was so bad you needed sunglasses, but the sweat raining down the lenses made them impossible to wear. One day Ed came back from lunch and said the bank thermometer showed it was over 100°F. He said if we were in the union they'd send us home. Then he peeled off his shirt and told me to hurry up.

I learned a lot from Ed, and I'm happy to say that most of what I learned can be found in this book. Not because Ed or I put it here, but because other builders from all over the country have graciously shared their knowledge with the readers of *Fine Homebuilding*. Reprinted here are 28 articles from past issues of the magazine. Collectively, these articles represent a graduate degree in framing.

—Kevin Ireton, editor

Economical Framing

To conserve materials and save money, avoid overbuilding and design on 4-ft. modules

by E. Lee Fisher

Back in January 1973, the National Association of Home Builders' (NAHB) Research Center was commissioned by the Department of Housing and Urban Development (HUD) to devise a cost-effective framing system. I was the director of industrial engineering on that project, and what came out of it was the Optimum Value Engineered (OVE) house.

OVE is like a streamlined version of conventional wood framing. NAHB took a long, hard look at waste—materials that added nothing to the strength, durability or marketability of stick-framed houses. The OVE house was built without superfluous materials. With less material to put into the house, there was less work to do, so money was saved on both materials and labor.

The OVE house proved to be safe, marketable and inexpensive: everything HUD wanted in 1973. Nowadays, with the environment getting almost as much attention as the economy, OVE makes even better sense. Because the system reduces the amount of lumber products in a house, it conserves natural resources.

Most builders, regardless of production volume and price of their homes, will find the OVE approach to framing easy to incorporate. Many builders throughout the country have already instituted many of the lumber- and plywood-savings methods prescribed by OVE. Most OVE methods are accepted by the major model building codes, but you should check with your local building officials before trying these methods.

Materials-based design—If you want to give OVE a try, start with design. A cost-effective floor plan has the most floor area enclosed by the least amount of exterior wall. For example, say you've got two homes, both of them with 1,200 sq. ft. of floor area. The first home has a 20-ft. by 60-ft. floor plan, so it's got 160 lineal ft. of exterior wall. The second home has a 30-ft. by 40-ft. plan; it's got 140 lineal ft. of exterior wall: 20 ft. less wall enclosing the same amount of floor area.

When designing a house, the goal should be to keep the floor-to-wall ratio as high as possible. In the example, the first home has a 7.5:1 floor-to-wall ratio; the second has a 8.5:1 ratio.

Lumber and sheet products—plywood, oriented strand board, particleboard, gypsum wallboard, etc.—are produced in 2-ft. increments. Minor adjustments on the plans can make more efficient use of lumber and sheet products, reducing waste and eliminating framing members.

Less material, equal quality. **Many builders are now using a modified version of conventional wood framing called OVE (optimum value engineered), which was developed by the NAHB Research Center to make houses more affordable.**

The most cost-effective design places all exterior walls and as many interior walls as possible at 4-ft. increments, or modules. This major module is divided into minor modules of 2 ft. (drawing facing page). This scale matches the width of most construction sheet materials and standard lumber lengths, thereby eliminating cutting labor and reducing scrap. Consider a wall that's 22 ft. 9 in. long. This wall takes just as much plate lumber and sheathing to build as a 24-ft. long wall (remember, lumber products are produced in 2-ft. increments). However, the shorter wall creates almost 4 ft. of scrap plate lumber and over 9 sq. ft. of scrap sheathing. In addition, at least one stud bay will be narrow, making the insulation contractor cut a batt to fit. The longer wall actually costs less to build.

Traditionally, roofs have been designed without much consideration for sizes of framing members and plywood sheathing. Plans are drawn for 4-in-12, 5-in-12 and 6-in-12 pitches without weighing the effect on efficient use of lumber and plywood. But there's nothing magical about whole-number roof pitches. Consider drawing a roof-sheathing layout that takes maximum advantage of the 4-ft. by 8-ft. dimensions of plywood and let the pitch fall where it may. This roof design eliminates plywood and lumber scrap plus the associated labor for cutting.

Designing around stud spacing—Window and door placement offer more opportunities for material savings during design. Placing at least one side of each opening so that it falls on a stud (photo above) eliminates up to half the studs used for framing openings, and sheathing can be placed without excessive cutting and fitting. Moving windows and doors only 2 in. or 3 in. can save as many as 20 studs per house, depending on the number of openings. Well-placed openings also reduce the number of very narrow stud bays, which are difficult to insulate.

Moving openings to fall on studs won't normally damage the house aesthetically, except if you move a large opening—a full bay window taking up nearly an entire wall, for example—where the small walls on both sides of the opening could be of obviously different widths.

In conventional wood framing, most builders use three studs or two studs and three blocks to form a partition post, or channel, where walls intersect other than at corners. But by designing interior walls to intersect exterior walls on 2-ft. modules, ends of the interior walls will hit exterior wall studs (OVE employs 2-ft. o. c. spacing), eliminating extra nailer studs and blocking (left drawing, p. 10). When an interior wall butts into an exterior wall stud, you just need drywall clips, not blocking, to fasten drywall. And again, by placing interior walls to fall on normally occurring studs, insulation installation is easier. As far as interior trim is concerned, eliminating corner studs doesn't make it tough for carpenters to nail baseboard at a corner. Most baseboard can just be nailed to the bottom plate.

Sometimes the 2-ft. modular design isn't possible. Bathrooms, for instance, are normally 5 ft. wide to allow room for bathtubs. When a stud isn't available to catch an interior wall, the next best bet is to use a horizontal block (right drawing, p. 10). The block is nailed in the exterior wall between regular studs 4 ft. from the floor to catch the drywall edge. The abutting wall is toenailed to the top and bottom plates. Then it's spiked to the horizontal block, which is nailed on the flat to allow room for insulation. Horizontal blocks can easily eliminate another 12 to 15 studs in an average-size house.

When the width of an interior wall is 4½ in., it's easy to see that the 2-ft. modular design won't eliminate the need to cut and fit drywall. OVE concentrates on saving exterior materials, which are more expensive than interior materials.

Overbuilt floors—Often, more wood products are used in floors than in any other part of the house. Floors in a 2,400-sq. ft. house can contain as much as 4,800 bd. ft. of lumber and as much as 4,800 sq. ft. of subflooring. Therefore, if you want to cut down on lumber and plywood use, pay close attention to the floors.

Consider eliminating all lumber and plywood on the first floor by switching to concrete slab-on-grade construction. Many builders will probably say they can't get away with concrete slabs in their markets. Buyers won't accept them. That's what Maryland builders said until Levitt and Sons moved into the state and built over 10,000 concrete-slab homes. One of the most successful builders of quality homes in the Cleveland area builds slab houses. Much of the large Chicago suburb of Hoffman Estates consists of slab homes. About half the homes built in the United States over the last 20 years are on concrete slabs. Slab-on-grade construction is worth considering because it can eliminate almost one-fourth of all lumber and up to one-third of the sheet wood products in a home. Because homes built on slabs have no basements, lost storage and workspace can be made up inexpensively with a larger garage or the addition of a storage room.

An easy way to reduce materials in floors is to eliminate midspan cross bridging. In the past, when wet lumber was often used, cross bridging helped keep joists from twisting as they dried. But tests on dry lumber conducted by the NAHB Research Center during the 1960s proved that cross bridging adds nothing to floor stiffness. Because of these tests, all major model codes consider cross bridging unnecessary for floors framed with lumber up to and including 2x12s.

Something that does add stiffness to the floor is structural adhesive. In some cases (depending

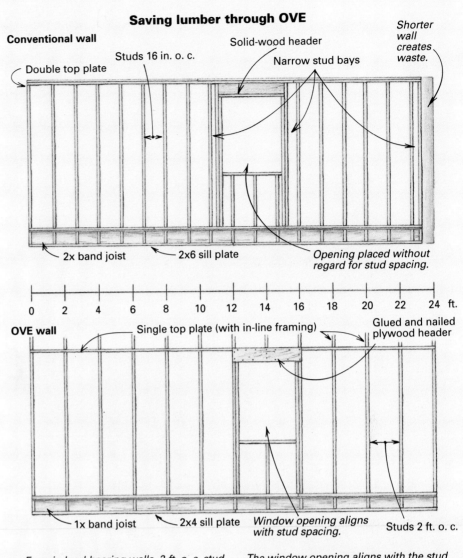

Saving lumber through OVE

Conventional wall

Double top plate · Studs 16 in. o. c. · Solid-wood header · Narrow stud bays · *Shorter wall creates waste.*

2x band joist · 2x6 sill plate · *Opening placed without regard for stud spacing.*

0 2 4 6 8 10 12 14 16 18 20 22 24 ft.

OVE wall

Single top plate (with in-line framing) · Glued and nailed plywood header

1x band joist · 2x4 sill plate · *Window opening aligns with stud spacing.* · Studs 2 ft. o. c.

Even in load-bearing walls, 2-ft. o. c. stud spacing is possible. When rafters and floor joists align with studs (in-line framing), the second top plate is omitted. A glued and nailed plywood header saves lumber, won't shrink and provides space for insulation.

The window opening aligns with the stud spacing, eliminating extra framing and narrow stud bays. And the 24-ft. wall uses the same plate and sheathing material as the 22-ft. 9-in. wall but requires no cutting and produces no waste.

on the quality of the floor components and on the loads the floor supports), gluing and nailing a subfloor can reduce the size or grade of floor joists needed or increase joist spacing from 16 in. o. c. to 19.2 in. o. c., or even 2 ft. o. c. The technique involves applying a structural adhesive (on top of joists), preferably an elastomeric adhesive, which remains pliable and fills voids, then nailing or screwing down the floor deck.

Most builders use a 2x6 sill plate on the foundation wall. Although it takes more precise placement of anchors, a 2x4 sill plate provides adequate bearing for the ends of floor joists.

The primary function of band joists is to hold the floor joists erect until floor and wall sheathing are installed. The band joist normally doesn't need to be the same thickness as the regular joists because exterior-wall loads are transferred to the floor joists. Both 1x stock and ⅝-in. plywood are adequate for band joists.

Some builders install double floor joists under all interior walls that run parallel to the floor framing. Double joists aren't necessary under non-load-bearing walls, which typically weigh less than the furniture that will go in a room.

For load-bearing walls parallel to floor joists, span the joist bay with blocks 2 ft. o. c.; if the wall bears on a joist, slap on another floor joist.

Lumber prices have just about caught up with those of engineered wood products, so consider substituting products, such as manufactured wood I-beam joists and other laminated-veneer lumber (LVL), for dimensional lumber. The quality of engineered wood products is consistently high, yet it takes less wood to make them, and the wood can be of lesser quality, so by using engineered wood products you're conserving higher-grade lumber.

Reducing excess lumber in walls—When combined, exterior walls and interior walls contain about the same board footage of lumber as the floors. Therefore, to frame economically, it makes sense to use wall lumber wisely.

When your design is based on a 2-ft. module, it makes sense to use 2-ft. spacing between framing

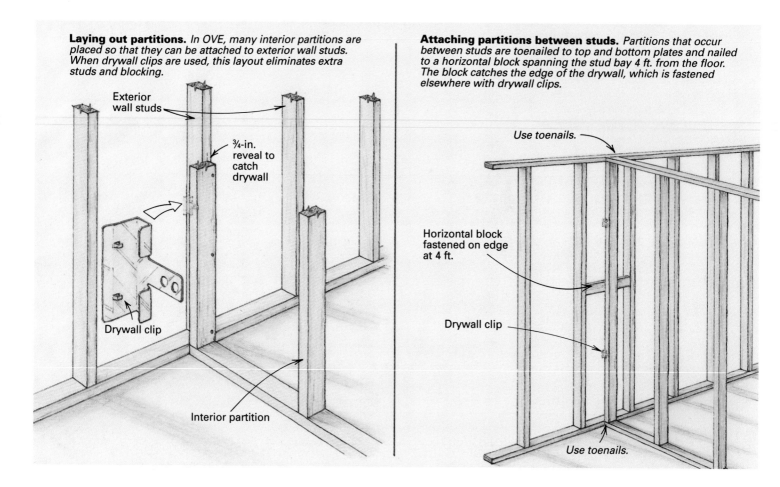

Laying out partitions. *In OVE, many interior partitions are placed so that they can be attached to exterior wall studs. When drywall clips are used, this layout eliminates extra studs and blocking.*

Exterior wall studs

¾-in. reveal to catch drywall

Drywall clip

Interior partition

Attaching partitions between studs. *Partitions that occur between studs are toenailed to top and bottom plates and nailed to a horizontal block spanning the stud bay 4 ft. from the floor. The block catches the edge of the drywall, which is fastened elsewhere with drywall clips.*

Use toenails.

Horizontal block fastened on edge at 4 ft.

Drywall clip

Use toenails.

members. The major model building codes allow 2-ft. o. c. stud spacing under certain conditions, even in load-bearing walls. For example, the Uniform Building Code permits 2-ft. o. c. stud spacing in one-story dwellings and on the top floor of multistory dwellings, which reduces framing material. Even if you are hesitant to place load-bearing studs on 2-ft. centers, at least consider doing it in non-load-bearing walls.

Even in a relatively small house, it's not unusual to have 40 to 50 corners in the walls. Typically, each of these corners is built with three studs and three blocks (top photos, facing page). The third stud is installed as a drywall backup strip, and the three blocks act as spacers for the third stud. By using drywall clips, one stud and all the blocking can be eliminated (bottom left photo, facing page). Alternatively, a plywood cleat screwed onto the inside face of a corner stud also provides nailing or screwing surface for drywall (bottom right photo, facing page).

Drywall clips come in different styles and materials, but the ones I've used (Prest-On, 316 Lookout Point, Hot Springs, Ark. 71913; 501-525-4683) slide onto the edge of drywall, then the clip is fastened to the corner stud. Just remember which piece of drywall goes on first; with Prest-On clips, the first sheet butts into the leading edge of the corner stud, and the second sheet butts into the first. The drywall itself shouldn't be fastened to the corner stud. A fastened corner is more likely to crack as the stud shrinks. Instead, fasten the drywall one stud away from the corner (2 ft. in an OVE system).

Two-stud corners can save as many as 40 to 50 studs and 120 to 150 blocks in a house with-

out affecting the structural integrity of the walls. Besides, with three studs and blocking, insulating is difficult and often skipped in the exterior corners, causing a number of 4½-in. wide cold spots.

If 2-ft. o. c. stud spacing is used in load-bearing walls, and walls in a home are designed to 2-ft. modules, framing layout is a snap; there are no odd dimensions to worry about. If floor and roof framing are also on 2-ft. centers, an opportunity for in-line framing exists. In-line framing, or stacking, means that rafters are located directly over wall studs, and wall studs are in turn located directly over floor joists. In-line framing directly transfers the load through the structure, which is much more efficient than the common zigzag route of load transfer.

For example, if in-line framing is used, the second top plate, or doubler, can be eliminated because its sole function is to help transfer loads to the nearest studs. To get a full 8-ft. ceiling height without using a doubler, studs must be 1½ in. longer than precuts. Precuts usually are priced at 8 ft., so the longer stud shouldn't cost any more than the old precuts.

Plywood headers—Headers are needed over windows and doors in load-bearing walls. However, many builders install solid or built-up wood headers over non-load-bearing openings, too, which is a waste of material and labor. Unnecessary headers are found not only in interior but also exterior walls, especially gable-end walls. Where headers are needed, an alternative to solid or built-up wood is a plywood box-beam header. Rough openings are framed as though

the wall were non-load-bearing, and then ½-in. Group I exterior-grade plywood is glued and nailed above the rough opening—on one side of openings up to 3 ft. wide that carry no more than 500 lb. per lineal ft., and on both sides of openings up to 6 ft. wide that carry the same 500-lb. per lineal ft. maximum load.

A good structural adhesive and a nailing pattern engineered to handle the load are required. If you install an interior plywood header, use ½-in. AC plywood with the A side (smooth side) facing the room; then tape and finish it with joint compound to match the drywall.

Solid-wood headers shrink, resulting in cracks in the drywall. But a plywood header is stable and won't cause cracks around the drywall. More importantly, plywood headers provide cavities for insulation.

Hollow-core interior doors in non-load-bearing walls are lightweight and do not need a trimmer stud, or jack stud, on each side of the opening. It is still best to double the stud on the hinge side of heavier, solid-wood doors.

It's not shoddy construction—The OVE approach is not a substandard building method. It gets the most performance from the least amount of material. Some builders have said that OVE requires them to supervise workers more closely. True, OVE requires more discipline because it's not as forgiving of labor and material defects as traditional approaches. On the other hand, OVE is a simpler way of building that creates less opportunity for defects to occur in the first place.

The OVE approach is not an all-or-nothing system. You can pick and choose the parts of the

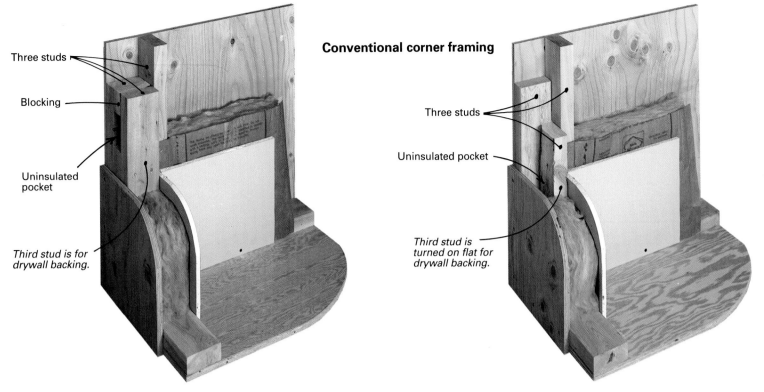

Conventional corner framing

Three studs

Blocking

Uninsulated pocket

Third stud is for drywall backing.

Three studs

Uninsulated pocket

Third stud is turned on flat for drywall backing.

Eliminating a stud *Conventional wood framing uses three studs per corner. The third stud is either built out with blocks or turned on the flat to provide drywall backing. An OVE corner requires only two studs, with drywall backing provided by drywall clips or a plywood cleat. Open and accessible stud bays in OVE corners are easy to insulate.*

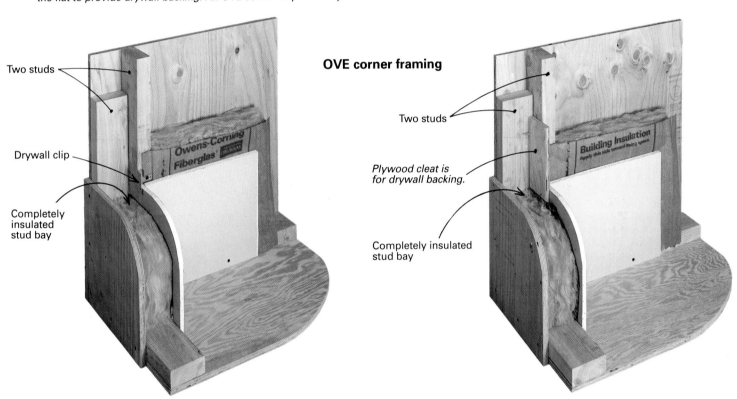

OVE corner framing

Two studs

Drywall clip

Completely insulated stud bay

Two studs

Plywood cleat is for drywall backing.

Completely insulated stud bay

system that make sense to you. Say, for example, that you want to build two-stud corners, but your clients are concerned about wavy walls. I personally have never seen a wavy wall that was caused by 2-ft. o. c. studs, not even with drywall ceilings hung on 2-ft. o. c. rafters. Two-stud corners work just as well with 16-in. o. c. framing as with 2-ft. o. c. framing.

Maybe your drywall contractor wants to charge extra to work with drywall clips. You know that drywall clips are cheap and easy to use, but you can't convince your contractor. The heck with it. At least you can save on studs by using 2-ft. centers. I'm not suggesting that blocking be elim-

inated where it's needed, so make sure that things like handrails, closet poles and towel bars have the proper backing.

You can begin eliminating headers over non-load-bearing openings tomorrow without changing anything else. Or you can start gluing and nailing subfloors because it eliminates labor by reducing callbacks. Saving materials and labor without sacrificing building quality is what OVE is all about. □

E. Lee Fisher is senior industrial engineer at the NAHB Research Center in Washington Grove, Md. Photos by Susan Kahn except where noted.

Further reading

The methods described in this article are detailed in two publications: *The Lumber and Plywood Saving Manual* and *Reducing Home Building Costs with OVE Design and Construction* (the OVE manual), both available from the NAHB Research Center, 400 Prince George's Blvd., Upper Marlboro, Md. 20772. The cost for each publication is $10 plus $2 for shipping and handling. An update of the OVE manual is currently in the works. *—E. L. F.*

Squaring and Leveling Mudsills

If you assume the foundation is accurate, you may end up custom cutting each rafter and fussing with every miter in your trim

by Rick Arnold and Mike Guertin

Sill width determines baseline. After a rough check to verify the overall dimensions of the foundation, a baseline is established by measuring in the width of the sill from the edge of the concrete.

Parallel walls are lined up from the baseline. The two crew members in the background mark the same measurement from the baseline on the wall behind them. The crew member in the foreground moves the chalkline until it lines up with the marks determining the sill location for the garage wall.

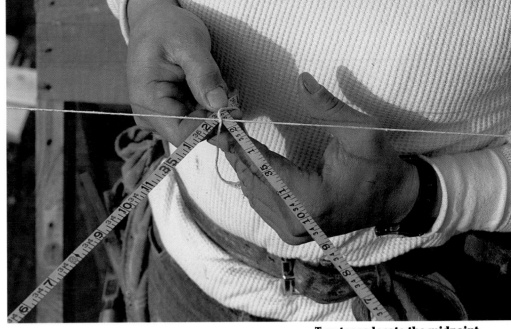

O n one of our first framing jobs, the lead carpenter installed the mudsills by just lining them up with the outside edge of the foundation. It wasn't until the first-floor deck was framed and sheathed that we realized the foundation was 3 in. wider at one end, which made the whole platform miserably out of square. We battled problems from this little oversight all the way through the roof, and soon after we began looking for a new lead carpenter.

In the 15 or so years since that project, we've become a little fanatical about squaring and leveling our mudsills regardless of the scale or price range of the house. The reward for being finicky is a first-floor platform that is square, level and built to the exact dimensions called for on the plans. This extra care that is taken at the beginning saves us time and headaches throughout the project.

Crew members must communicate at all times to get accurate measurements— Measuring foundations is almost always a two-person job and may even involve three people, if more than one measuring tape is being used at one time. Constant communication between all crew members involved in a measurement is of utmost importance to the success of squaring and leveling.

We have an unwritten rule that the person on the beginning, or dummy, end of the measuring tape always calls out the measurement he is holding to the line as well as the color of that line—"5⅝, blue line," for instance. The person on the business end of the measuring tape always acknowledges his partner's call before recording the measurement.

To maintain the highest level of consistency and accuracy, we also insist that crew members never switch ends of the tape or switch jobs with anyone else on the crew during the entire process. On past projects we wasted hours trying to figure out why dimensions didn't jibe only to discover that a crew member merely held the

Two tapes locate the midpoint. A crew member crosses two tapes that are being stretched from the ends of the baseline. A small piece of twine is tied on the dry string where the two tapes meet at exactly the same length. This point marks the center of the rear wall.

Dry strings are used where the foundation drops. To guarantee square mudsills on the dropped foundation, the crew projects the measurements to dry strings held at the same level as the rest of the foundation. The homemade device for holding the dry strings has an angle iron that is nailed into the foundation and a telescoping pipe to allow the crew to adjust the height of the strings.

Diagonal measurements are the final check for square. After the corners are located on the rear wall, the crew measures between them and the baseline corners. If the measuring procedures have been followed carefully, the two diagonal measurements should be within ⅛ in.

Leveling the foundation begins with a reference point. A crew member makes a mark 1½ in. down on one corner of the foundation. From this point the crew uses a water level (photo facing page) and marks every corner. A level line between the marks will help locate places in the foundation that might be out of level.

A plumb bob transfers the corner point to the wall below. The intersection of the two dry strings is the exact corner of the sill layout. That point, as well as any others needed to snap chalklines, is found by dropping a plumb line from the dry string.

tape in the wrong spot and that his partner failed to check him.

Check the foundation right away—We always square the foundation and snap lines for our mudsills right after the forms have been stripped (and before the foundation contractor is paid), even if we know that we won't begin framing for a while. Although totally unworkable foundations are rare, we have seen some foundation pours that were out of level as much as 3 in. in 40 ft. and some foundations that varied up to 12 in. from the specifications on the plan. Our crew can work comfortably with foundation tolerances that are within ¾ in. of square and ½ in. of level, and overall dimensions that are within ½ in. of the foundation plan.

We also double-check the foundation plan against the first-floor plan before lines are snapped. Simple features such as shelves for exterior brick veneer can have a way of throwing

off foundation dimensions. Any deviation from the plan or discrepancy that exceeds our tolerances adds cost to the project and needs to be addressed before we begin. The foundation that we will use to illustrate our methods is 40 ft. long and 30 ft. wide with an 8-ft. by 20-ft. breezeway that connects the house to a 24-ft. by 24-ft. garage. The rear foundation wall of the house is dropped 7 ft., and the gable-end foundation wall steps down with the grade.

After inspecting the foundation for any obvious problems, such as a large bump-out or a severe crown or dip, a couple of crew members check the overall foundation measurements. At the same time other crew members sweep dirt or concrete debris off the top of the walls.

We install a double-2x6 mudsill system common to this part of southern New England, using pressure-treated lumber for the lower sill. With the double-mudsill system, leveling shims are inserted between the two layers, giving us a nice,

A water level is quicker, easier to use and more accurate than a transit. Having tested many kinds of leveling devices, including builder's transits, this crew's tool of choice is a water level. The tool requires no calibration and holds up well under the rigors of house framing.

tight seal against the concrete while providing a flat surface for the first-floor deck. We always check the width of a few pieces of the pressure-treated stock to determine the exact distance to set our chalkline back from the edge of the concrete. Most often this measurement is 5⅝ in.

Having inspected our foundation and found that it is close to the plan specifications, we choose the longest grade-level wall and mark 5⅝ in. from the outside of the concrete on each end (left photo, p. 12). We snap a line between these marks in blue chalk to serve as a baseline.

We snap our lines in blue chalk initially. Then, if the line needs to be adjusted for any reason, we switch to red. The crew has been instructed that a red line always supersedes a blue one.

The rear sills are lined up parallel to the front—The next step is establishing a rear-wall line parallel to our baseline. Foundations that are poured all on one level are the easiest and quickest to deal with because lines and measurements can be made directly onto the foundation. One crew member holds the dummy end of the tape at 5⅝ in. over one end of the baseline while the other member with the business end of the tape marks the planned width of the house minus 5⅝ in. on the rear wall. The process is repeated at the other end, and a line is snapped on the rear wall between these marks.

As mentioned before, the rear wall of this particular foundation is 7 ft. lower than the front wall, and the end wall is stepped, which complicates the procedure slightly. We've devised a fixture that is nailed to the lower foundation corner and that holds a dry string (in place of the chalkline) at the height of the grade-level portion of the foundation (center photo, p. 13). Instead of marking and snapping a chalkline on

the rear foundation wall, we run a dry string at the same elevation as the front wall. This string is positioned 29 ft. 6⅜ in. from the baseline (30 ft. for the width of the house minus 5⅝ in. for the width of the sill). Because the rear wall of the garage happens to be in line with the rear of the house foundation, we extend the dry string all the way through to the gable end of the garage and drive a nail into the concrete there to anchor the string in position. This dry string will remain in place as a reference line until we've finished installing the sills.

We establish the parallel line for the front wall of the garage by measuring 6 ft. from our baseline on the two walls that run perpendicular to the front wall (right photo, p. 12). We extend a chalkline from the 6-ft. mark on the end wall of the house across the front wall of the garage. The line is then moved until it aligns with the 6-ft. mark on the intermediate perpendicular wall. Before the line is snapped, we secure it and verify that it is parallel to the dry line at the rear wall of the garage by measuring between the two. If our measurements are off slightly, we tweak the line at the end of the garage to compensate. When we're satisfied, we snap the line for the garage wall in blue. We can now locate and snap any other parallel lines, such as the front wall of the breezeway, by referencing one of the snapped lines or the dry string.

Perpendicular lines are located by simple geometry—We have just established the lines for our front and rear walls. The distance between these lines is the length of our perpendicular walls. Next we locate the corner points that will give us the lengths of our front and rear walls as well as the placement of perpendicular walls, and we can begin to square the foundation.

When we made a rough measurement of the foundation earlier, we determined that the actual length of the front wall along our baseline was close to the specified length of 40 ft. So next we mark 5⅝ in. in from the outside of the concrete at one end of our baseline. One crew member holds the tape at 5⅝ in. on that mark while a second pulls the tape along the wall and marks the baseline at 39 ft. 6⅜ in., or 5⅝ in. back from the total length of the wall.

Our next step is establishing a perpendicular line for the gable end of the house exactly 90° to the baseline. In the past we tried the 3-4-5 right-triangle method, which got us close but not perfect. We used right-angle prisms, but we found them to be slow and hard to work with. The Pythagorean theorem also works fine, but the calculations have to be precise. Also, figuring in the 5⅝-in. setback on top of all that will befuddle even the most mathematically inclined. Our latest method has outperformed all others in simplicity and speed.

Three people + two 100-ft. tapes = one square foundation—There are actually two variations to our method. The first and most efficient method requires three people and two 100-ft. tapes.

Two crew members hold the 2-in. line of their tapes on the corner marks at each end of the front-wall baseline. We use the 2-in. mark because it is usually the first whole-inch mark on a 100-ft. tape; it also gives the crew member plenty to hold on to when the tape is being stretched tight. The third person holds the business end of both tapes near the middle of the line or, in this case, the dry string on the opposite wall. This person crosses the two tapes and moves left or right along the line until the measurements on the two tapes match exactly (top photo, p. 13). This point indicates the exact center of the rear wall, and it's marked either with a pencil line on the foundation or a piece of twine tied around the dry string (an alligator clip from an electronics-supply store also works well to mark a point on a dry string).

From the midpoint we measure half the length of the wall minus the width of the sill in each direction to get our inside-corner marks (center photo, p. 13). For insurance we double-check the overall length of the rear wall. Now, as a final check for square, we measure diagonally from corner to corner on the foundation (bottom photo, p. 13). If we've done our job properly, those measurements should be within 1/8 in.

Two crew members and one tape take a little longer—If there are only two crew members or just one 100-ft. tape available, we use a modified method that's a little slower but works just as well as the first system. We mark the length of our baseline just as before. Then one person holds the tape at 2 in. on one end of the line.

Instead of crossing two tapes, the guy on the business end stretches the single tape across the opposite parallel line near the midpoint and picks the closest 1-ft. increment on the tape. For our purposes we'll call that measurement 42 ft. The crew member on the business end then moves the tape along the line and marks where the 42-ft. mark on the tape intersects with the chalkline or dry string. The dummy end of the tape is then moved to the mark at the opposite end of the baseline, and the tape is held at the same 2-in. point. The business end again marks the point where the 42-ft. mark on the tape intersects with the line. The midpoint between these two marks should be the midpoint of the wall. Just as before, we measure to the left and to the right of this midpoint half the length of the wall, again minus the width of our sills, to locate our corner points, and again we confirm overall squareness with diagonal measurements.

With both methods, once we're sure that we have a perfect rectangle, we snap lines for the perpendicular walls between the corner points on our front and rear walls. From those lines, we can measure and snap any other lines that are perpendicular to our baseline, including the gable-end wall of the garage.

Both rear-corner points of this foundation happen to be "in the air" because of the drops in the foundation. On the corner where our dry-string fixture is mounted, we indicate the line of the gable-end wall using another dry string instead of a snapped line. The intersection of the two dry strings tied to our fixture is our exact corner point (left photo, p. 14). Starting at this point, we drop a plumb bob down to the lower foundation walls at enough locations to let us snap lines. But on a windy day even the tightest dry string will tend to move a bit. And even the slightest breeze can make using a plumb bob impractical. If the wind turns our plumb bob into a pendulum, we reluctantly use a 4-ft. level in its place.

Water levels are the most accurate tools for leveling sills—For years we used a builder's transit to set our sills level, and it worked fine

A simple jig locates the bolt holes. With the sill stock lined up on the inside of the chalkline, a jig with a U-shaped end is held against the bolt. The bolt-hole location is marked through a hole 5/8 in. from the U.

most of the time. We still use one occasionally if it is the best-suited tool. However, transits are delicate and vulnerable instruments, particularly around framing crews on job sites. Invariably, the transit gets knocked over or ends up bouncing around in the back of someone's pickup. Consequently, our transits seemed to spend more time in the shop being adjusted than they spent out in the field.

The tool we've come to depend on for leveling our mudsills is the good, old-fashioned, low-tech water level. (For more on building and using water levels, see *Fine Homebuilding* #85, pp. 58-60). We've been accused of being too cheap to buy a transit, and skeptics doubt the accuracy of our water-filled plastic hose. But our water level consistently outperforms the transit for both speed and accuracy. In fact, we have used our water level to inform the foundation contractor that his transit needs calibration.

We first pick a corner of the foundation and mark 1 1/2 in. down from the top of the concrete (right photo, p. 14). After checking the plastic hose for air bubbles and making sure the fluid level in both ends is equal, one crew member, designated as the lead person, raises or lowers the hose until the water level matches his starting mark while another crew member on the other end of the hose marks the level of his corner at the lead's command (photo p. 15). For the sake of consistency, we always make our mark at the bottom of the meniscus (the concave shape of the water surface inside the hose).

We try to mark as many corners of the foundation as our length of hose allows from a single reference point. If a corner is beyond the reach of the hose, the lead person repositions himself at the farthest mark, and the process resumes.

On some foundations it is necessary for the lead person to move four or five times to catch all

Shims correct an out-of-level foundation wall. If there are dips in the foundation, the two sills are pried apart with a flat bar. The nails used to tack the upper sill in place now serve to hold the two sills apart while shims are inserted beneath the future joist layout.

the corners of the foundation. We mark the entire grade level of the foundation and always work our way back to our starting corner to check ourselves. We are usually within ⅛ in., which is an acceptable tolerance.

If there is a drop or a step in the foundation, we mark the level on the wall just before the drop. Next we measure the drop and round to the nearest inch. We then measure that distance down from the upper-level mark and make a new lower mark on the dropped wall. We continue the leveling process from this point, marking all the dropped foundation walls at each corner until the wall elevation returns to the original level. At this point the measurement between the upper and lower levels should be within ⅛ in. of the original drop measurement.

After making level marks around the entire foundation, we go back and snap lines between our marks. Sometimes we have to scrape off excess concrete at the form-panel seams if it interferes with the chalkline. After the level line is snapped, a crew member takes measurements from the top of the concrete to the chalkline at random points around the foundation. This process gives us an idea of any areas where we'll have to adjust the level of the sills.

Nuts on foundation bolts start out only finger-tight—Once all of our lines are snapped, we can install the sills. We cut the lower sill stock to length for each section of foundation and line it up on the inside of the chalkline (photo facing page). Next, we locate the bolt holes with a homemade tool made from a piece of metal with a U-shape on one end and a hole drilled ⅝ in.

back from the U. We butt the U against the bolt with the tool squared to the sill stock by eye and mark the bolt location through the hole on the other end of the tool. The holes we drill in the mudsills are oversize so that we can adjust the sills to the chalklines once they're in place.

We roll out sill seal just before dropping the sills over the bolts. The lower sill is held in place with masonry nails until the upper sill can be marked, drilled and set on top. We cut the upper 2x6 sills so that the corners cross lap the lower sills. We also try to overlap any butt seams in the upper and lower sills by at least 4 ft. to make straightening bowed sill stock an easier job. We avoid landing seams over windows or bulkhead openings. We extend upper sills beyond any foundation drops by at least 4 ft. to tie into the kneewall below.

In most cases the kiln-dried lumber we use for the upper sill is narrower than the pressure-treated sill on the bottom. In order to maintain the proper outside dimensions of the house framing, we keep the inside of the bottom sill flush with the chalkline on the foundation, and we keep the outside of the upper sill flush with outside of the lower sill. We tack the upper sill in place with a couple of nails every few feet, and we put the nuts and washers on the foundation bolts only finger-tight at first in case shimming is necessary.

Shims between the sills bring them up to level—Most of the foundations we work on are within ¼ in. from the highest point to the lowest point; and by the time we lay the sills on top of the sill seal, these differences are negligible. When we do encounter a more significant dip in the foundation, we use flat bars to pry the sills

apart, and we insert shims between the sills to bring them up to level (photo above). The nails we used to tack the two sills together now hold them apart while we slip in the shim shingles. We try to place the shims directly beneath the joist location for the best support of the platform.

If we find that an entire wall is low by ¼ in. or more, we rip 5½-in. strips of the appropriate thickness plywood and sandwich them between the two sills. Occasionally, we encounter a hump in a foundation. Rather than shim all of the sills up to this level of the hump, we make the adjustment later by scribing the rim joist and notching the joists slightly as long as the hump isn't huge. This procedure is preferable to shimming the sills around the entire perimeter of the foundation.

Once the shimming is complete, we tighten the nuts, but only snug so we don't create additional dips in the sills. We complete the installation by nailing the top and bottom sills together according to the local code.

As a final note, the entire process of squaring and leveling a foundation usually takes three crew members about two hours. It takes the same three guys another two hours to drill, shim and nail the sills in place. Expect these methods to take a bit longer at first, but with practice they'll become quicker and easier. Also, that level line snapped around the foundation will come in handy later on for orienting the siding. □

Rick Arnold and Mike Guertin are partners in Midcor Construction, a building company, as well as U. S. Building Concepts, a construction consulting business in East Greenwich, R. I. Photos by Roe A. Osborn.

Floor Framing

With production techniques and the right materials, a solid, squeakless floor is a day's work

by Don Dunkley

Rough framing isn't measured in thirty-seconds of an inch, and for good reason. Although a shabby frame can put finish carpenters in a murderous mood, it gets covered up and forgotten behind siding, drywall and paneling. But floors, if done poorly, will come back to haunt you. That squeak just outside the bedroom door is an annoyance for which there is no quick fix. But the problems can go beyond the floor itself. The blame for eaves that look like a roller coaster once the gutters are hung often rests squarely on a carelessly built floor two stories below.

Haste doesn't usually create these problems; using inappropriate materials and not knowing where to spend extra time does. In fact, by using production techniques and materials, rolling second-story joists (tipping them up and nailing them in place) and sheathing them with plywood on a modest-sized house is a day's work for my partner and me.

What a floor does—Most builders are very aware when they hang a door of the abuse it will get, but they don't think in the same way about the floor they are framing. This skin is the main horizontal plane the building relies on to transfer live loads (ones that are subject to change, like furniture and people) and dead loads (primarily the weight of the structure itself) to the bearing walls, beams and foundation. Critical here is the amount of deflection in both the sheathing and the joists. The subfloor must be stiff enough to handle both general and concentrated loading, and to support the finish floor that rests on it without too much movement. But at the same time it must be flexible enough for comfortable walking and standing, something a concrete slab can never be.

Floor systems—There are many different ways to build a floor, and the preference for one system over another has a strong regional flavor. Floor trusses—either metal or wood—are becoming popular because of their prefab economy, and because they can eliminate the need for bearing walls, beams and posts by free-spanning long distances. Girder systems are still used over crawl spaces in some areas. They are generally laid 4 ft. o. c. with their ends resting on the mudsill of the foundation wall, or in pockets cast into the wall itself, or on metal hangers attached to the foundation. The interior spans of the girders are typically posted down to concrete piers. Girders are decked with either 2x T&G or very heavy plywood designed for this purpose.

But most wood-frame houses still use joists. And whether a floor rests on foundation walls on the first story or on stud walls on a higher level, its elements are pretty much the same. Joists are typically 2x lumber, laid out on 12-in., 16-in. or 24-in. centers and nailed on edge. They are held in place on their ends by toenails to the sill or plate below, and then attached to a perimeter joist, or blocked in between. Blocking or bridging has also been used traditionally to stabilize a floor at unsup-

ported midspans. (For more on this, see the next page.)

The last element of any joist system is the skin—usually plywood sheathing. This decking not only forms the continuous horizontal surface, but is also the key to the integrity and structural continuity of any wood floor. I'll talk more about the choice of material and how to apply it later.

In the simplest floor system, joists rest on the mudsill of exterior foundation walls, or on the double top plate of the stud walls on higher stories. With the long spans found in most floor plans, joists have to be either very deep (say, 12 in. or 14 in.), or supported at or near the center of their length by beams or bearing walls, creating two shorter spans. In the case of even greater building widths, joists can come from either side of the building, lapping each other over these supports, with blocking nailed in between.

In much of the country where basements are used to get below the frost line, the joists bear on or lap over built-up beams of 2xs spiked together. These are usually supported by Lally columns—concrete-filled tubular steel posts with flanges top and bottom—or by hefty wood posts.

In the West, where I live, the first-story floor usually stretches over a crawl space. Perimeter support for the joists is provided either by stemwalls—mudsill-capped low foundation walls (most crawl spaces use the minimum allowable height of 18 in. from grade to the bottom of the joists)—or by cripple walls (sometimes called pony walls) that are framed up on hillside foundations to reach the first-floor level.

For supporting the joists in mid-length or for lapping them, floors over crawl spaces either use a carrying beam, or girder, on 4x posts that rest on concrete pier blocks, or use a crib wall—a low, framed wall studded up from an interior concrete footing that runs the length of the building.

Unlike girders, which are often crowned and leave a hump in the finish floor, crib walls can be plumbed and lined to make a perfectly straight, level surface. They will also shrink less than a large girder, and just feel a lot more solid to me. The posts under a girder are toenailed at the bottom to the wooden block that caps each concrete pier. This block inevitably splits, making the whole post-and-beam assembly feel a little cobbled together. I've also learned that concrete contractors get bored with straight lines at the point when they begin laying out piers, so supporting posts often get precious little bearing.

Bolting down the sill—Floors need to be level, square and precise in their dimensions. On a second floor, these conditions will depend on your accurately plumbing and lining the first-floor walls; but on the first floor, setting the mudsill is the critical step. First, check the perimeter foundation walls with a steel tape for the dimensions shown on the plans. Then pull diagonals to check for square. Run 3-4-5 checks on the corners for

square if the floor isn't a simple rectangle. Once you're sure of your measurements, snap chalklines that represent the mudsill's inside edge on the top of the foundation walls.

Be thinking ahead to the sheathing (called shear panel where I build) and siding if the foundation walls aren't arrow-straight. If these are one and the same, the plywood will need to hang down over the concrete foundation wall an inch or two. This means that you're better off having the mudsill overhang the foundation slightly, rather than the other way around. If finish siding is used over sheathing, this isn't as important, since the sheathing can butt the top of the foundation where it bows out from the mudsill.

Place the mudsill (either pressure-treated or foundation-grade redwood) around the perimeter of the foundation, and cut it to the lengths needed to fit on the stemwalls. You'll find foundation bolts placed on maximum 6-ft. centers; code requires each piece of sill to be held by at least two bolts. The cut sills should be laid in position on their edges along the top of the stemwalls so that they can be marked for where the anchor bolts fall along their length.

After carefully squaring these marks across the face of the sill, lay them flat so you can mark the bolt positions across their width. Measure the distance from the chalkline on the top of the stemwalls to the center of each bolt. Then transfer that measurement to the face of the mudsill, taking care to measure from the inside edge of the mudsill each time.

Production carpenters who do a lot of lay-out use a bolt marker. One of these can be made from an old combination-square blade, as shown in the drawing below. The end of the metal blade is notched so that it will index from the centerline of a ½-in. foundation bolt. At 5½ in. from this end for a 2x6 mudsill (or 3½ in. for a 2x4), a hole is drilled in the blade and fitted with a screw or nail.

To use this marking gauge, the mudsill has

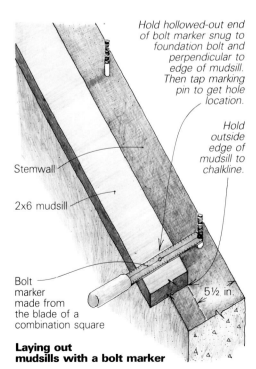

Hold hollowed-out end of bolt marker snug to foundation bolt and perpendicular to edge of mudsill. Then tap marking pin to get hole location.

Hold outside edge of mudsill to chalkline.

Stemwall

2x6 mudsill

Bolt marker made from the blade of a combination square

5½ in.

Laying out mudsills with a bolt marker

Bridging midspans

Although the days when all carpenters learned the trade through a formal apprenticeship are gone, construction knowledge is still passed down through the ranks, from those who know to those who want to. This kind of conservatism is often valuable, but it also means that outmoded methods die hard. Bridging between joists at midspan is a good example. Even on relatively short spans, a lot of knowledgeable builders are convinced that bridging guarantees a stiffer floor. But convincing studies show that it doesn't, and that information isn't new.

Bridging is a continuous line of bracing that runs perpendicular to the direction of the joists, and is installed between them. It can take the form of cross bridging (diagonal pairs of wood or metal braces that form an X in the joist space) or horizontal bridging (continuous, full-depth solid blocking).

A study done by the National Association of Home Builders Research Institute Laboratory in 1961 put most of the myths to rest. It proved with laboratory and field tests that bridging at midspan had little effect in stiffening a floor once it was sheathed. It was also of little help in transferring concentrated loads laterally to other joists, reducing floor vibration or preventing joists from warping. Bridging actually does all of those things as you install it,

but once the plywood subfloor is on, its net effect is negligible.

Still the myth persists. Building in the 1970s, I put in midspan solid blocking every 8 ft. on floors I built, to meet the code and because I was concerned with quality building. But there is only one way in which midspan bridging improves a floor substantially, and that is in providing lateral stability. Because joists are on edge, they tend to tip unless they are pinned in at their ends and at support points like girders and bearing walls. Plywood sheathing nailed and glued on top also helps. But over long spans, the joists can use a bit more help.

Of the three model codes in the U. S., only ICBO's Uniform Building Code requires using cross bridging or solid blocking with plywood sheathing under normal conditions, and that is only when the depth-to-thickness ratio is 6 or more, as with 2x12s. It is then required every 8 ft. along the span of the joists. BOCA requires bridging or blocking only if the joist depth exceeds 12 in. under normal loading conditions, and SBCC allows you to skip it entirely in single-family residential buildings.

If you do need to install midspan bridging or blocking on a floor you're building, a few simple tricks will make it go faster and eliminate some of the squeaks. First, you must decide between diagonal bridging, which can either be wood 1x3s or the newer sheet-metal bracing, and solid blocking.

I think that cutting and nailing the old wooden cross bridging is needlessly time-consuming, although all the standard construction texts cover it in detail. If I were going to use bridging, I would choose the metal variety that doesn't use nails. These straps have sharp prongs on their ends, and eliminate the squeaking you get from nails rubbing on the metal bridging when the joists flex underfoot. This kind of bridging can be installed from above or below the joists; a combination of the two is even better. Drive the upper ends of the bridging into the top of each joist face before sheathing, and return to drive the bottoms in from below when the joists have done most of their shrinking, and initial loading has flattened them some. Make sure to separate each member of the pairs to prevent contact squeaking.

Solid blocking has to be done just before sheathing, but it's straightforward. If you're going to insulate the floor yourself, it's the best choice. To begin, chalk a line across the joists where the blocking goes. This is typically at the center of any span under 16 ft., or in two rows equally spaced if the span is greater. Alternate the blocks on either side of this line so you can face-nail them through the joists.

Set a plank on top of the joists, parallel to the chalkline and about a foot away. Set it to the left of the line if you're right-handed. Spread out your blocks along this plank. Drive three 16d nails partway into the face of each joist down the line, alternating between the right and left side of the line. This allows you to position each block and then anchor it with a single blow of the hammer. Once you've traveled the entire length of the joists nailing the blocks on one end, you can turn around and travel back the way you came, nailing the other ends.

Construction adhesive recommended for solid blocking is supposed to cut down on squeaks, but it's messy and time-consuming. It's more practical to drive shims in any gaps that open up between the blocks and the joists. Don't use vinyl sinkers (vinyl-coated nails) or brights (uncoated steel nails). These will loosen up and squeak much sooner than ring-shanks, screw nails or hot-dipped galvanized.   —*Paul Spring*

to be held flat with its outside edge on the chalkline. This is most easily done from inside the foundation. Make sure that the sill is exactly where you want it along its length. Then by holding the bolt marker perpendicular to the length of the sill, with the notch of the gauge pressing against the bolt, tap on the marking pin with your hammer. Whatever the location of the bolt, the correct bore center will be left on the sill.

After the sills are laid out for bolts and drilled with a slightly oversize bit (a 9/16-in. bit for 1/2-in. bolts is a good compromise between requirements for a snug fit and giving the carpenter a break), place them on the stemwall and hand-tighten the nuts. If termite shields or sill sealers (wide strips of compressible insulation that act as a gasket to keep air from passing between foundation and framing) are to be used, be sure to lay them in underneath the mudsill. Now is also the time to adjust the foundation for level by shimming under the mudsill where necessary with grout. Go back and cinch the nuts down tight once everything is set.

With a hillside foundation where the joists will rest on a cripple or pony wall, adjustments for level are made in the cripple studs themselves rather than by shimming the plate. If the foundation is stepped down the hill, these cripple studs will be in groups of various lengths. If a grade beam that parallels the slope is used, then the cripples will have to be cut like gable-end studs. In this case the beveled end will be nailed to the mudsill and blocked in between. In either case, getting the top plates precisely level (and straight) is worth the trouble. Do this by shooting the tops of the studs with a transit, or by setting up a string-and-batterboard system. Using a spirit level for long runs just isn't accurate enough, and trusting a hillside foundation to be true is a mistake you'll make only once.

Joist layout—The first thing I do to lay out the sills for rolling joists is to get out the plans again and find all the exceptions to the standard layout. These include stair openings that have to be headed off, lowered joists for a thick tile-on-mortar finished floor, plumbing runs that need to stub up on a joist layout, and the double joists used for extra support. Even a simple floor will require doublers to pick up the weight of the building at its most concentrated points.

The first place to double is at each end of the foundation, parallel to the direction the joists run. When joisting a second story, sandwich short 2x spacer blocks between two joists for a tripler (drawing, facing page). This will give you backing for the ceiling drywall below. Locate all interior walls running parallel to the joists and lay out the sills (and carrying beams if you have them) for doublers too. Since some plumbers feel that joists have just as much integrity cut in half as they do whole, it's a good idea to have a chat with your pipe bender before you complete the joist layout. This will save a lot of headaches later. For instance, the doublers under plumb-

Heading off the joists to make room for DWV lines, as shown here, is a lot easier if plumbing runs are considered before framing begins.

ing walls will have to be spread for waste lines that run parallel to the joists, or headed off with solid blocking to accommodate perpendicular plumbing runs (photo above). Doubling joists under a heavy cast-iron bathtub is also a good idea.

Next, lay out the joist spacing shown on your plans. This layout has to be adjusted so that the butting edges of the decking will fall in the center of a joist, allowing two sheets to join over a single joist. Stretch your tape from

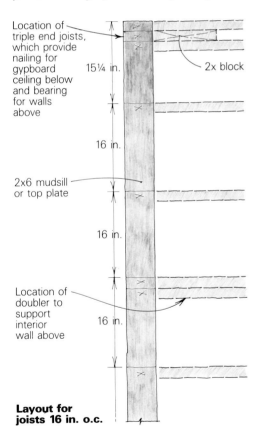

Location of triple end joists, which provide nailing for gypboard ceiling below and bearing for walls above

15¼ in.

2x block

16 in.

2x6 mudsill or top plate

16 in.

Location of doubler to support interior wall above

16 in.

Layout for joists 16 in. o.c.

the outside edge of the sill and make a mark ¾ in. (or half the thickness of a 2x joist) shy of the joist-spacing dimension—15¼ for 16 in. o. c. Put your X on the leading side (far side) of this mark for the first joist. The rest of the layout can be taken from this first joist at the required centers without any further adjustment. If your floor has joists coming from each side that lap over a beam or wall, the layout on one side will need to be offset from

the layout on the other perimeter sill or plate by 1½ in. to account for the lap.

After you have completed laying out the sills and midspan beams, check again to make sure that they jibe, and that the centers are accurate for the plywood. Now is the time to think; once you begin rolling joists, you just want to move.

Unlike rim joists, the end-blocking method requires cutting special-length blocks to correct for layout and accommodate double joists.

Getting ready to roll—There are two ways to secure the ends of the joists. The first is to run them the full width of the house, and block in between at their ends (photo above). This system has some structural advantages since the joist ends are locked in, but with all of those short, brittle pieces involved, it requires more fussing to get straight lines. The other way is to cut the joists 1½ in. short on each end, and run a 2x rim joist (also called a header joist, ribbon or band joist) perpendicular to the joists and face-nail it to them.

Whichever system you use, all the joists should be laid flat on the sills right next to their layout marks before you begin to frame. Stocking is basically a one-man job. As you pull each one up, sight it for crown. Many framers mark the crown with an arrow (make sure it points up), but it's not necessary if you start at one end of the building, and lay each joist down so that the crowned edge is leading. This way, when you return to the beginning point, you'll be able to reach for the leading edge of each one and tilt it up or roll it, leaving the crown up. Save the really straight lengths for rim joists. If any of the joists are badly crowned, cut them up for blocking.

Rim joists—The rim joist for one side of the building should be installed before any of the joists. Mark its top edge with the same layout as the sill it will sit on. If a large crown holds the rim up off the top plate or mudsill, cut the rim joist in half on a joist layout mark. Also, if this floor is over a crawl space, you will need to cut in vent openings. Place these every 8 ft., starting 2 ft. in from the corners, where air tends not to circulate. Typical vent openings are 5 in. high and 14½ in. long; these fit nicely between joists on 16-in. centers.

Install the rim joist with toenails from the outside face of the board down through the sill or top plate every 16 in. Once nailed, string the outside edge to make sure it is

straight and adjust it in and out if necessary with a few more toenails. Now snug up the joists that are resting flat on the plate or sill so they butt the inside face of the installed rim joist (this is easily done by catching the face of the joist that's up with the claw of your rip hammer and pulling toward the outside of the building, as shown below). If any joist ends are badly out of square, trim them.

Using the perimeter mudsills or plates, which have already been carefully plumbed and lined, as a kind of tape measure, mark

The author snugs a joist against the rim to check the end for square, using the building itself as a ruler for cutting the joists to length.

Kiln-dried framing makes better joists, but in the West green Douglas fir is much more common. Even so, the ideal moisture content of a joist is 15%. Many of the problems with floors—in particular squeaking and nail-pop—can be traced to wet joists shrinking away from nail shanks as the wood dries. Wet joists can permanently deflect once they begin to dry, causing a swale in the floor. Most of the so-called settling in a platform-framed house is due to the joists shrinking across the grain. Much of this can be eliminated by using dry lumber. Putting down a plastic moisture barrier over the earth in damp crawl spaces is a good way of keeping the floor above dry.

Cutting to length. The joists are squared up 1½ in. short of the building line (below) to leave room for the rim joist, and cut to length. With their crowns marked by an arrow, they are ready to be tipped up.

Rolling the joists. Joisting (right) is fast if a routine is established between partners. On the open side of the joists, an end toenail holds the joist on layout; toenailing the joist face to the plate will follow. On the other end, the rim joist is face-nailed to the joist with three 16d nails, then toenailed to the plate. The end joist in the foreground will be doubled later.

Installing the other rim. With joists nailed off, the open side gets its rim (bottom). This floor uses 2x14s; 2x10s are more typical.

each joist where it extends over the perimeter wall opposite the installed-joist rim. Square up a line 1½ in. in from this outside mark, and cut. For a floor with more than one span, index off the outside of the girder or crib wall, and use the rule of thumb for lapped joists to determine if they will need to be cut. The joists are now ready for rolling.

With the rim joist installed, rolling goes fast, as the joists can be quickly tipped up and nailed in place—no special blocks to slow you down and no need to stretch a tape measure to make sure you're keeping the layout, as in end blocking. Rolling joists (photo facing page, top right) is a synchronized, almost fluid process that requires few words once partners are used to working together. With one of you on each side of the building, and beginning with the doublers at one end or the other, reach forward at the same time for the top of the joist, tip it up and place both ends of the joist on their layout marks. On the side of the building where the rim is already installed, drive three face nails through the rim into the end grain of the joist. If a joist is a little lower in height than the rim when tipped up, hold it up flush to the top edge and shim

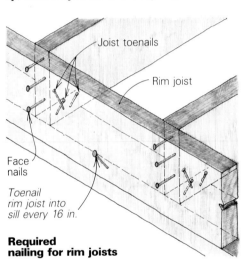

Joist toenails

Rim joist

Face nails

Toenail rim joist into sill every 16 in.

Required nailing for rim joists

up after the joists are rolled. The other member of your crew on the opposite side will drive a single toenail through the end of the joist, down near the bottom, into the top plate or mudsill. Then, both of you drive two toenails through the leading face of the joist down into the plate or mudsill (the side facing the joists that haven't yet been tipped up), as shown above. Once you've rolled all the joists, you can reverse direction, and work your way back toenailing the other face of the joists to the plate or mudsill. This is called backnailing. Then the rim on the open side of the joists should be hauled into place, toenailed to the plate or sill, and face-nailed to each joist (photo facing page, bottom).

End-blocking method—If you choose end blocking over a rim-joist system, each joist will need a block, which should be precut and stacked on top of its joist as it lies flat and ready to be rolled. A radial-arm saw set up with a stop gauge is the easiest way to gang-cut blocks, but a skillsaw will do. For a 16-in.

layout, the blocks should measure 14⁷⁄₁₆ in. The more accurate your blocks, the fewer specials you'll have to cut while you're joisting. The first block, however, will be a special to account for the outside doubler and the ¾-in. adjustment that was made in the layout for the sheathing. After all blocks are cut, put them on the joists that are ready to be rolled. Proceed to roll the joists first by setting up the outside doubler. If you have to use more than one piece to make the length, then make sure that the butted splices are at least 4 ft. from each other for these double joists.

Once the doubler is in place, toenail the first special block to the doubler and to the sill or plate—first with a toenail through the

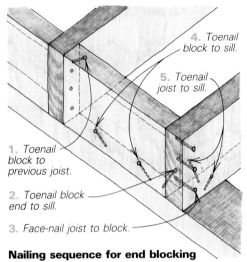

4. Toenail block to sill.

5. Toenail joist to sill.

1. Toenail block to previous joist.

2. Toenail block end to sill.

3. Face-nail joist to block.

Nailing sequence for end blocking

top of the joist into the block it butts against; second, with a low toenail into the sill or plate through the end grain of the block on the leading end. Then tip up the next joist in front of you and face-nail through it into the end grain of the block with three nails. Finally, drive two toenails from the outside of the block into the sill or plate (drawing, above) and toenail the joist itself down. Tip up all joists in the same manner. Although the plate is laid out, it is a good idea to measure joists every 4 ft. or so to make sure they aren't wandering off the layout. Shorten a block or cut a longer one to adjust if needed. This will take a lot of frustration out of the sheathing later on.

Lapped joists—In most floor plans, the distance to be spanned is long enough to require that a girder or crib wall be used to support the joists, or as a bearing surface where two runs can lap. Double-check the height of this girder or wall to avoid the typical center hump that a lot of floors develop after a year or so. To tell the truth, this support is best set slightly too low than too high. Crib walls will shrink less than most beams, but using dry lumber will help in both cases.

Joists can be butted and a plywood or lumber splice-plate at least 24 in. long used to join them, but most often they are lapped. The rule of thumb I've always used is that the joists should lap each other by at least 4 in., but that neither one should extend beyond the far edge of the girder or crib wall that sup-

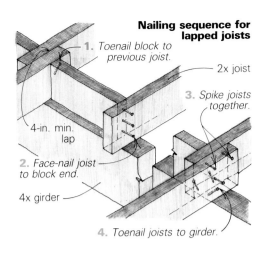

Nailing sequence for lapped joists

1. Toenail block to previous joist.

2x joist

3. Spike joists together.

4-in. min. lap

2. Face-nail joist to block end.

4x girder

4. Toenail joists to girder.

ports it by more than this joist's nominal depth. The joists should be toenailed to the support beam or plate through each face, and then blocked solid in between, as shown above. These blocks have to be shorter than end blocks to make up for the extra joist. Pay attention to the layout by pulling a measurement from the end of the building every 4 ft. to make sure that the joists aren't bowed in or out from the layout on the perimeter, and that the subfloor joints will fall on a joist. (Remember, you may be dealing with two layouts. If the joists lap, the layout on one side of the floor will differ from the other by 1½ in.)

Schedule the rough plumbing after joisting is completed and before you begin the sheathing. If you are using forced-air perimeter floor heat, the ducts should be installed as well. Be sure to block around the heat registers in the floor to provide backing for the sheathing once the subs have finished the rough-in. Then call for inspection and get it signed off. This is also the time for insulation.

Sheathing—Joist systems can use either a single or double layer of sheathing. The traditional approach is to sheathe the joists first with subfloor—typically ½-in. or ⅝-in. square-edge plywood laid with its face grain (and long dimension) perpendicular to the joists. This forms a working platform for rough construction, but is covered over with underlayment (either plywood or dense particleboard) ranging from ⅜ in. to ¾ in. This produces a thick floor with little flex, and more resistance to sound transmission so the patter of little feet is muffled in rooms below.

However, most of the houses I build use a single-skin, combination subfloor and underlayment. It requires half the labor and produces a serviceable floor. These plywood panels are tongue-and-grooved on their long sides (the ends will always fall on a joist for continuous support), and are structurally rated to span the joist spacing by themselves. It's possible to order square-edge panels, but with these, continuous edge-blocking between joists is required. Single-skin panels range in thickness from ¹⁹⁄₃₂ in. to 1⅛ in. (known as 2-4-1). Always use the thickest you can afford. While code allows ⅝-in. plywood on 20-in. centers, and some builders are already calling for its use with glue on 24-in. centers; ¾-in. plywood on 16-in. centers has my vote. Inch-

Sheathing the floor. The first step in sheathing is to establish a course line 4 ft. in. from the rim with a snapline (top). Installing T&G plywood using a 2x4 beating block and a sledgehammer is a job for two. The author, in the rear of the photo above, is using both his shifting weight and his hammer to coax the tongue into the groove. This second course begins with a half sheet in order to stagger the joints. On this job, white glue was specified; construction adhesive is far superior.

and-an-eighth plywood (2-4-1) is even stiffer, and can be used on 16-in. or 24-in. centers even though it was designed for use over girders placed at 4 ft.

As in underlayment plywood, the voids in the face veneers and the plies directly beneath them in combination subfloor and underlayment are plugged so that the wood resists point loading (consider what high heels can do to plywood with interior voids). The panels are also drum-sanded to a precise thickness. With this plywood, about the only place a second layer is used around here is with vinyl flooring in baths and kitchens. Underlayment is particularly important with thin flooring because it provides a new dense surface that hasn't taken the abuse of construction traffic. Installed just before the finish flooring, it lessens the likelihood of joint lines or nailheads broadcasting through flooring. This extra layer also brings the level of these finished floors flush, or nearly so, with areas of the house that may have padding and carpet, tile or hardwood flooring.

Setting plywood—Check T&G plywood when it's delivered. If either long side is banged up, crushed by banding, or looks like it's been sitting out in the rain, don't even let them unload it. The tongue is difficult enough to thread into the groove without sabotage. Once you've got the material on site, treat it with care. Cover it even if clear skies are predicted.

To sheathe the floor using plywood, first snap a line 4 ft. in from one side of the building, using a level to plumb up from the mudsill or top plate of the outside wall below for accuracy. The butt joints of plywood have to be staggered between courses (there should never be an intersection of four corners), so you should begin either the first or second course with a half sheet. Remember that you are laying the plywood down with its length running perpendicular to the direction the joists run. Begin the first course with the tongue sitting on the rim joist at the perimeter of the building. This will leave the groove as the leading edge. The next course of plywood will have to be threaded onto the previous course by driving the new sheets with a 2x4 block and a sledge hammer. The grooved edge will take this abuse; the tongue won't.

If you have lapped joists, the layout for the sheathing will be different from one side of the floor to the other by 1½ in. If the lap occurs at a natural break for the plywood, the next course can just be slid back 1½ in. so that it continues to butt on a joist. But if a plywood course covers the joist lap, then you'll have to spike 2x4 scabs to each joist to get bearing for the plywood at this transition.

Gluing down the plywood with an elastomeric construction adhesive in addition to nailing enables the plywood and joists to act as one integral unit, creating what amounts to a T-beam. Using glue will lessen creep and squeaking and increase floor stiffness. It also lets you cut down the nailing schedule on plywood ¾ in. thick or less from 6 in. on the edges and 10 in. in the field, to 12 in. for both

using 6d (although I prefer 8d) screw nails or ring-shank nails. Construction adhesive comes in tubes that are used with a caulking gun. Using large tubes saves time, and buying by the case will save you a few bucks. White glue is better than nothing, but it won't bond as well as the neoprene-base adhesives do.

Apply a bead of adhesive continuously on every joist, but don't work too far ahead of yourself because it will skin over rapidly. Flop the plywood down carefully so that it's very close to where it belongs. On the first course, aim for the chalkline, and then nail off the sheet with the exception of the last 6 in. of width along the groove. Nail this little bit once the next course is driven into place or you will pinch the groove and make threading the tongue nearly impossible.

Plywood associations and textbook writers agree that this kind of combination subfloor and underlayment should be spaced ⅛ in. at edges and ends. The climate where I build is very dry, and this practice isn't common, but I can see that you might need to allow for expansion in some climates. If you do use this recommended spacing, you will need to trim panels for length occasionally since you will be adding slightly to the layout each time. Or you can scab on a 2x4 to the existing joist to get more bearing. This technique is particularly useful at the end of a course of plywood when the floor is just slightly longer than the last full sheet will stretch. Although plywood should always span at least two joist spaces, the exception is where a small piece rests completely on solid joists such as a double or triple. If your layout falls just short of the end of the floor, you can add a scab to the end joists to support the last full piece of plywood, and fill in with a ripping.

After the first course of plywood is nailed off, you can begin the next course, remembering to stagger the joints. This is best done with three people. The first lays down the bead of adhesive ahead of the plywood and drops back occasionally to nail sheets that have been tacked before the adhesive sets up. The other two will be cutting and setting the plywood. The only trick to setting is a bit of coordination between the person handling the sledge on the leading edge of the new panel, and the person using his feet and body weight to thread the seam (photo bottom left).

Once set, each sheet should be nailed at all four corners and at each joist, which will help whoever is nailing off keep the nailing lines straight. Nail off the plywood with ring-shank or screw nails because they hold much better than conventional nails, particularly in slightly green joists. Make sure the nails go straight into the joists. Nails that break through the sides will cause squeaks. It's worth sending someone below to drive out any that miss.

Once all sheathing is nailed off, trim off overhanging edges where necessary with a circular saw. Your efforts will result in a good strong subfloor that will withstand all the wear and tear a household will give it. □

Don Dunkley lives in Cool, Calif.

Laying out for Framing

A production method for translating the blueprints to the wall plates

by Jud Peake

On big production projects where jobs are highly specialized, the carpenter assigned to layout uses a hammer very little. His tools are a lumber crayon (known in the trade as keel), a pencil, a layout stick, a channel marker and a tape measure. If he marks the top and bottom plates correctly, the carpenters may not even need a tape measure to frame up the walls. All of the figuring that involves the plans—and there are a lot of variables to anticipate—is done at the layout stage.

On large jobs, I don't necessarily frame the house I lay out. On houses I contract, I do both jobs. Either way, I treat layout as if there won't be anyone around to answer questions when the framing starts. Doing the layout as a separate operation increases the speed and accuracy of framing, whether you're building your own house by yourself or working as part of a big crew. The plans are an abstraction of the building to be constructed. Layout systematically translates your blueprints into a full-size set of templates—the top and bottom plate of each interior and exterior wall on a given level of the house. The pieces of the puzzle can then be cut and framed in sections.

There is little about a finished house that isn't determined by the layout. The actual procedure falls into four distinct steps. Within each step, you have to deal with wall heights

and the locations for windows, doors, corners, partitions, beams and point loads. You also have to deal with *specials* (which is anything else and usually means prefab components).

The first step in layout is to go over the blueprints and mark them up with the information you'll need, in the form it will be most useful. Step two is to measure out the slab or deck and establish chalklines representing every wall on that level. This is called *snapping out.* In step three you'll decide where walls begin and end and which ones will be framed first, and then cut top and bottom plates for each wall. This is called *plating.* Finally you *detail* the plates by marking them with all the information the framer will need to know to build the walls. Layout requires a thorough knowledge of framing. But even then, regional framing techniques vary widely.

Layout principles—Layout is based on parallel lines. If two lines are parallel and one is plumb, then the other will be plumb. Also, if a pair of lines meet at a right angle, then another pair of lines, each parallel to its counterpart in the first pair, will meet at a right angle. Stated less theoretically, put a 2x4 on top of another 2x4 and cut them to the same length. Using a square, draw a line across the edges of both every 16 in. Frame in between

these top and bottom plates with studs of equal length, and stand the wall up and plumb one end. Now all of the studs will be plumb. Doors and windows that have been laid out with these studs will be plumb. And if your foundation and deck are square and level, then any interior walls paralleled off the outside walls will also be square and plumb.

The second principle is to do all the layout at once. This way all the plates are fitted against each other, and you can be confident that the walls will work together before they grow vertically and become unmanageable.

The third principle is my own: avoid math. You can do most of your figuring in place. If you're looking for the plate and stud lengths of a rake wall (one whose top is built at the pitch of the roof it supports), snap it out full scale on the deck. Measure things as few times as possible. Once the walls are mapped out on the deck in chalk, don't measure them and then transfer this number to the plate stock sitting on sawhorses. Instead, lay the plate material right on the line and cut it in place. This saves time and reduces error—the kind that has to be fixed with a cat's paw.

Marking up the plans—Before you step out on the slab or deck, you need to go over the blueprints and systematically pick out the ele-

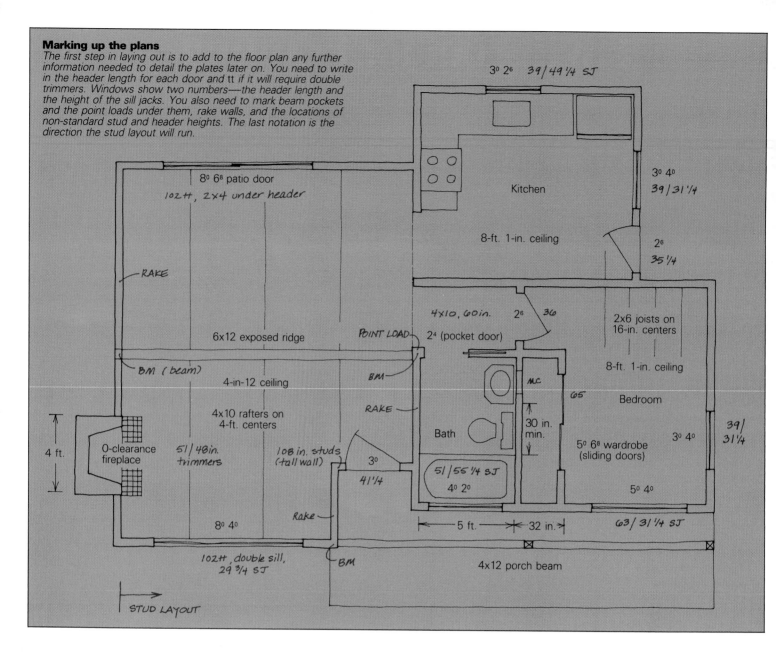

Marking up the plans

The first step in laying out is to add to the floor plan any further information needed to detail the plates later on. You need to write in the header length for each door and tt if it will require double trimmers. Windows show two numbers—the header length and the height of the sill jacks. You also need to mark beam pockets and the point loads under them, rake walls, and the locations of non-standard stud and header heights. The last notation is the direction the stud layout will run.

3⁰ 2⁶   39/49¼ SJ

Kitchen

3⁰ 4⁰
39/31¼

8⁰ 6⁸ patio door
102 H, 2x4 under header

8-ft. 1-in. ceiling

2⁶
35¼

RAKE

4x10, 60in.   2⁶   3⁶

2x6 joists on
16-in. centers

6x12 exposed ridge

POINT LOAD   2⁴ (pocket door)

8-ft. 1-in. ceiling

BM (beam)

4-in-12 ceiling

BM

RAKE

Bedroom

M.C.

65

4x10 rafters on
4-ft. centers

4 ft.

0-clearance
fireplace

51/48in.
trimmers

108 in. studs
(tall wall)

3⁰

Bath

30 in.
min.

5⁰ 6⁸ wardrobe
(sliding doors)

3⁰ 4⁰

39/
31¼

41¼

51/55¼ SJ
4⁰ 2⁰

5⁰ 4⁰

8⁰ 4⁰   Rake

8⁰ 4⁰

102 H, double sill,
29¾ SJ

BM

◄—— 5 ft. ——►│◄ 32 in. ►│

63/31¼ SJ

4x12 porch beam

STUD LAYOUT

ments that should be part of the layout. I do this in the evening when I can slow down my pace a little bit. Thumb through the drawings just to review the general structure. Read the rough and finish carpentry specifications, and underline anything that isn't standard.

Next, go back to the first-floor plan in the blueprints. Write on the plans, next to the appropriate opening, the length of each door and window header. This will allow you to measure out and mark them on the plates later on without having to stop and figure, and will also allow whoever does the framing to make up a list and cut all of the header stock, sill and sill jacks at once. See sidebar on p. 27 for what these framing members are, and how to figure their lengths.

The drawing above shows a blueprint that contains many of the framing situations you will encounter. I'll use these same plans in explaining snapping out, plating and detailing. In this case, the blueprint is marked up with the necessary information for laying out. For doors, only the header length is written in; when nothing else is noted, this indicates a standard 6-ft. 8-in. door in an 8-ft. wall.

In the case of the aluminum patio door in

the living room, a 2x4 is needed to fur the header down to the right height. The *tt* next to the header length indicates double trimmers for this 8-ft. opening. The pocket door to the bathroom shows the length of the header, and its narrower width (4x10). Windows show two numbers: the first is the length of the header, and the second is the length of the sill jacks.

With the exception of the rake walls, which will be snapped out full scale on the deck, I have marked the only wall that doesn't use standard 8-ft. studs (the jog in the living room) with the notation *108-in. studs (tall wall)*. With a hodgepodge of elevations, it helps to mark up the plan view with colored pencil to differentiate stud heights.

The plans have also been marked for beam pockets *(BM)* on both ends of the ridge beam in the living room, and where the porch beam bears on the exterior wall. Resulting point loads (where significant loads have to be carried down to the ground) also have to be marked. The interior end of the ridge beam might not be picked up below without a notation. Also, studs should be doubled under joists that will be doubled on the floor above.

You'll also need to mark the dimensions of

any prefabricated items that will go into the house, like a medicine cabinet or roof trusses. (When using trusses, the building width can always err slightly on the narrow side, but don't ever make it too wide.)

One of the last things you'll want to note on the plans is at which end of the building you'll start the regular stud layout. The decision is yours. The side with the fewest jogs or offsets in the exterior walls is usually the place to start. When you reach a jog, compensate with the stud layout so that the studs, joists and rafters (or trusses) all line up, or *stack*. If the plans call for 2x6s on 2-ft. centers with trusses, code requires that they stack. It's a good idea anyway, both for increased structural strength, and for making multi-story mechanical runs like heat ducts easier.

In the drawing, I decided to pull my layout from the bottom left corner. By beginning there, the studs will pick up the rafter layout in the living room (exposed 4x10s on 4-ft. centers) and will work well with the ceiling joists and the porch overhang. You can also write down the cumulative dimensions of rooms, so that when you're snapping out, you'll be able to mark all of the intersecting

Illustrations: Frances Ashforth

The Pieces to the Puzzle
An introduction to the components of a modern frame and how to size them

If you've done some framing, it takes only a second of looking at the detailing on a set of plates to know what the wall will look like when it's framed and raised. After that it takes only a pencil and a 2x4 scrap to make up a cut list of headers, sill jacks, rough sills, trimmers, channels and corners. But if you're new to how all of this goes together, you'll need to understand the different framing styles and the many exceptions created by using different kinds of windows, doors and finish.

Below I've explained how the basics work. Since I learned my framing in the West, my explanation will focus on how it's done there using basic production techniques, but I'll also point out more traditional methods as I go along.

Stud walls—Studs (2x4s or 2x6s on 16-in. or 24-in. centers) hold up the roof or floor above, and provide nailing surfaces at regular intervals for the interior and exterior finish. Precut studs are 92¼ in. long and are used to build a standardized 8-ft. wall that works economically with standard plywood and drywall sizes. The actual height of this wall is 8 ft. ¾ in. once you add 1½ in. each for the *bottom plate (sole plate)* that sits under the studs, and the two *top plates* (the *top plate* and the second top plate, called the *double top plate* or *doubler*) that complete the wall. Architects sometimes add to the confusion by specifying this same wall as 8 ft. 1 in. Wall studs are seldom shorter, but where economy isn't as important, they are often longer. If walls are over 10 ft., they will require *fire stops*—horizontal blocking that slows down the upward spread of fire.

To find stud height, check the elevations and sections in the blueprints. What will be listed here are wall heights, usually shown as finished floor to finished floor (*F.F. to F.F.*). In most cases, you can use the same dimension for rough floor to rough floor. To figure the stud length given this dimension, subtract 4½ in. (the thickness of the three plates), plus the subfloor thickness and joist depth.

Rake walls—These are also called gable-end walls, and they require the framing to fill in right up to the bottom of the pitched roof. This means that each stud will be a different length and will be cut at the roof pitch on top. Typically, a rafter will sit on the top of these walls. There are two ways to frame them. One is to build the lower part of the wall just as you do the walls under the eaves of the roof and then fill in the gables later. This is fine if there is a flat ceiling at the 8-ft. height. The other way is to build the wall in one unit with continuous studs. This is necessary for cathedral ceilings. See p. 29 for the specifics on figuring lengths and angles.

Wall intersections—When one wall intersects another, the framing has to provide a solid nailed connection between the two walls, and backing for the interior finish. The drawing above left shows how this is usually handled. Money can be saved by replacing backing studs with nail-on drywall corner clips, and by reducing corner units to a simple L, but there are disadvantages to these cutbacks.

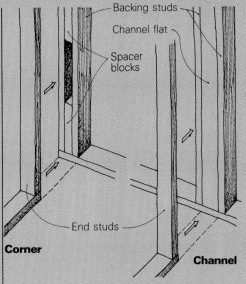

Corner

Backing studs
Channel flat
Spacer blocks
End studs

Channel

Doors—Door openings require *headers* to shift the weight of the roof in that area to both sides of the door. The vertical 2x support at each end of a header is called a *trimmer* (or cripple); the stud just outside of the trimmer that nails to the end of the header is called a *king stud*. The *rough opening* is the *rough-opening width*, measured between the trimmers, by the *rough-opening height*, which is measured from the floor to the bottom of the header.

Finding the height of the trimmers and the length of the header requires working backwards from the dimensions of the finished door. But you don't have to go through the full process each time—door headers should be 5 in. longer than the nominal width of the door. For example, a 37-in. header is needed for a 2-ft. 8-in. door. This 5-in. increment assumes that the trimmers will be framed very nearly plumb. Some carpenters use 5⅛ in. or even 5¼ in. to allow for sloppier framing.

Adding 5 in. to the header accommodates two 2x trimmers (1½ in. apiece, for a total of 3 in.) under the ends of the headers (drawing, above right). This leaves a rough opening 2 in. wider than the door. The remaining room is for two 1x side jambs (¾ in. apiece, for a total of 1½ in.), and ½ in. for shim space. (This leaves ¼ in. on each side.) Allow another ½ in. for exterior doors whose rabbeted jambs are closer to ⅞ in. thick. French doors will require closer to ½ in. of extra header length to account for the astragal (the vertical trim between the two doors that acts as a closure strip). If the door opening is wider than 8 ft., then the trimmers will need doubling, which requires another 3 in. of header length.

In the West, it's common to use 4x12 Douglas fir header stock in all 8-ft. walls. This system is fast, because all the framer has to do is cut the stock to length and nail it to the top plate—there are no *head jacks* to toenail, and you end up with a header at the right height that will span almost any opening. This system is admittedly wasteful, but gets around the cost of labor. Fir is still relatively inexpensive in large hunks, and the labor to cut and install the head jacks isn't.

If you aren't using 4x12s, check a span table for the correct header size (unless you are dealing with a non-bearing interior wall, where two flat 2x4s will do nicely). Typically, when

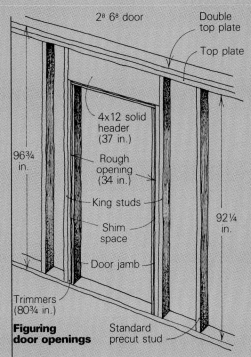

2⁸ 6⁸ door
Double top plate
Top plate
4x12 solid header (37 in.)
96¾ in.
Rough opening (34 in.)
King studs
Shim space
Door jamb
92¼ in.
Trimmers (80¾ in.)

Figuring door openings

Standard precut stud

solid headers aren't used in 2x4 bearing walls, the choice is a laminated header made on site from two lengths of 2x with ⅜-in. plywood sandwiched between. Using either system, if a wall exceeds 96¾ in. in height, head jacks (cripples) will be needed between the header and the top plate.

Using a standard 6-ft. 8-in. door, the trimmers should be cut 80¾ in., no matter how tall the wall is, or what kind of header you use. Once the bottom plate is cut out within the doorway, this will leave a rough-opening height of 6 ft. 10¼ in. This height will accommodate the door (6 ft. 8 in.), the *head jamb* (¾ in.), and enough play for the finish floor and door swing. An aluminum patio door requires 1½ in. of furring under the header. Pocket doors and some bifold doors require an extra 2 in. of rough-opening height for their overhead tracks. In an 8-ft. wall, a 4x10 held tight to the top plate works nicely. Even if you are not using solid header stock, you'll need trimmers that are 82¾ in. for these doors.

Windows—A rough window frame is like a door opening with the bottom filled in—that's just how you frame one. The rough width of a window is measured between the trimmers. The rough height is measured from the bottom of the header to the top of the *rough sill.* This is a flat 2x, doubled if the window is 8 ft. or wider, that runs between the inside faces of the trimmers. Unless it's otherwise noted on the plans, windows are framed with the same height trimmers as the doors.

In some areas of the country, the trimmers are installed in two pieces (*split trimmer*, or *split jack*) with the rough sill cut 3 in. longer and sandwiched between. If double trimmers are used, then the inside pair can be framed this way. Underneath the rough sill, the stud layout is kept by *sill jacks* (or cripples), which are, in essence, short studs.

Finding the length of the rough sills and the sill jacks is fairly simple. Depending on how you deal with the trimmers, the length of the rough sill will be the same as the width of the

Figuring rough window openings

Wood windows used to be specified by lite sizes such as the double-hung 32/22 (two lites, each 32 in. wide by 22 in. high) at right. But these days, you are more likely to get a unit size like 37¾ in. by 53¼ in. (which includes sash and jamb allowances), or better yet, the rough opening of 38¼ in. by 53½ in. (which also includes the space for shimming). These numbers will allow you to figure the header, sill and sill-jack lengths so that the window will fit.

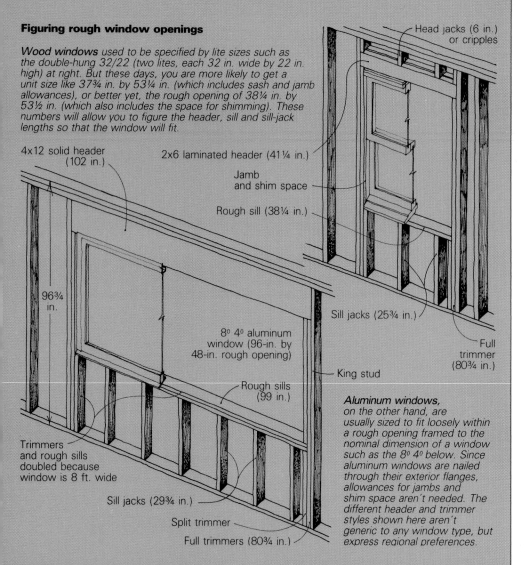

- Head jacks (6 in.) or cripples
- 4x12 solid header (102 in.)
- 2x6 laminated header (41¼ in.)
- Jamb and shim space
- Rough sill (38¼ in.)
- 96¾ in.
- Sill jacks (25¾ in.)
- 8⁰ 4⁰ aluminum window (96-in. by 48-in. rough opening)
- Full trimmer (80¾ in.)
- King stud
- Rough sills (99 in.)
- Trimmers and rough sills doubled because window is 8 ft. wide
- Sill jacks (29¾ in.)
- Split trimmer
- Full trimmers (80¾ in.)

Aluminum windows, on the other hand, are usually sized to fit loosely within a rough opening framed to the nominal dimension of a window such as the 8⁰ 4⁰ below. Since aluminum windows are nailed through their exterior flanges, allowances for jambs and shim space aren't needed. The different header and trimmer styles shown here aren't generic to any window type, but express regional preferences.

rough opening or 3 in. longer. To get the length of the sill jacks, add the rough opening height to the thickness of the rough sill, and subtract this from the height of the trimmer.

The length of a window header has a lot to do with what kind of window will fill the hole—aluminum or wood. To get the header length on an aluminum window less than 8 ft. wide, just add 3 in. to the nominal window width to allow for a trimmer on each side (drawing, above). This is because aluminum windows are generally manufactured to fit loosely in a rough opening of their nominal size. Check with your supplier to be safe. The first number stated for windows (and for doors) is the width— a 3⁰ x 5⁰ aluminum window needs a rough opening 36 in. wide and 60 in. high. For windows wider than 8 ft., you'll need to use double trimmers, which will increase the length of your header 3 in.

Wood windows were once specified by the size of the lite, or glass, which didn't account for the sash or frame that surrounded it. The rules of thumb about how many inches to add for the sash to get the rough-opening size are quite general because the width of stiles and rails differs among manufacturers. The rules are also specific to window type (double-hung, single or double casement, sliders, awning and hopper) since the number and width of their stiles and rails vary.

Today, most manufacturers' literature gives a unit dimension (this includes sash and frame) and a rough opening dimension. Use the rough-opening measurements for laying out, and add 3 in. to get the correct header

length on windows under 8 ft.; 6 in. over 8 ft. If the rough-opening size isn't specified in the plans or by the manufacturer, you've got to measure the windows on site. If they are set in jambs, add 3½ in. (two trimmers plus ½ in. for play) to get the correct header length; if not, add 5 in., as you would for a door.

Specials—This catchall category includes any member or fixture in the frame that's big enough to worry about. Medicine cabinets and intersecting beams are *specials*. Toilet-paper holders aren't; they can be dealt with later.

Your main concern with most specials is to provide backing for the interior wall finish, and a break in the regular stud layout if an opening is required. Beam pockets are simple. Imagine a ridge beam that is supported on one end by a rake wall. There are a number of ways to frame the bearing pocket for the beam, but any of these schemes should include a post or double stud beneath the beam, and backing on either side of it for nailing wall finish or trim.

Tubs, showers and prefab medicine cabinets require both backing and blocking. Blocking is what it sounds like—short horizontal blocks or a 2x let into the studs for nailing the wall finish. Backing is usually an extra stud and does the same thing only vertically. In the case of a corner bathtub, then, the line of blocks that sits just above the lip of the tub where the drywall will be nailed is the blocking. At the end and side of the tub where the wallboard will butt, you'll find an extra stud for backing. —*Paul Spring*

Snapping out. Facing page: Peake uses his foot to anchor one end of the chalkline while snapping out a short interior partition. The red keel X by his right foot tells the framer on which side of the line to nail down the wall once it's raised. This project was unusual because the exterior walls were framed, sheathed and raised before the interior was laid out.

interior walls along one side of the building by pulling the tape from just one point rather than having to do this room by room.

Snapping out—Snapping out isn't much more difficult than redrawing the architect's floor plan full size on the deck. Using a chalkbox, you need to snap only one side of each wall and draw a big X with keel every few feet on the side of this line that the wall will sit.

Measure for the outside walls first. Come in 3½ in. (for a 2x4 wall) or 5½ in. (for a 2x6 wall) from the edge of the deck. Be sure to measure from the building line, not from the edge of the plywood, in case it's been cut short or long. Don't even trust the rim joists without checking them with a level for plumb, since they may be rolled in or out.

Snapping lines should go quickly. To hold the end of your string on a wood deck, just hook it over the edge of the plywood, or use a nail or scratch awl driven into the deck. If the slab is very green (poured less than a week before), a drywall nail will usually penetrate the concrete; if not, use a concrete nail. You can even hold the string with one foot if the wall that you're snapping is short (see photo on facing page).

On very long walls, especially on windy days, have someone put a finger or foot near the middle of the line and snap each side separately. This will keep the chalkline truer. If you're working alone, close the return crank on the chalkbox to lock it and hook it over the edge of the deck so it's secure on both ends.

I use red chalk for layout. Lampblack also shows up well, but blue isn't a good choice in my area because it's the favorite of plywood crews. When I expect rain or even heavy dew, I use concrete pigment (Dowmans Cement and Mortar Colors, Box 2857, Long Beach, Calif. 90801) instead of chalk. If you're doing a lot of layout, get yourself a couple of chalkboxes with gear-driven rewind; if you've made a lot of mistakes, correct them with a different color chalk. When it comes to keel, I use red mostly; blue has a way of disappearing.

In snapping out the exterior walls, don't be concerned about intersecting lines where the walls come together. This problem will be solved when you do the plating. Just concentrate on getting the lines down accurately. Once you've done that, check the lines for square by measuring diagonals, and check the dimensions again carefully. On slabs and first floors, check to see that where you have put the exterior walls will allow the siding to lap over the concrete for weathertightness.

I usually snap out rake walls on the deck (see sidebar, p. 29). This will let the framer cut the studs in place between the chalklines, once again avoiding the math. Make sure that you

Rake walls

Rake walls are fairly simple to plate and detail if you snap them out full scale on the deck, and keep in mind how they will relate to the rafters that will eventually sit on them. The bottom plate of the wall will look like any other. The trick is locating the top plate so the framer can fill in the rake-wall studs without doing any calculations.

In this case the rake wall will intersect a standard 96¾-in. wall. The first real complication is dealing with the bird's mouth on the rafter. With 2x4 walls, I use a 3½-in. level cut so that the rake wall dies into the 8-ft. wall at the top inside edge of its double plate. This also allows me to measure the run of the rake wall from the inside of the 8-ft. wall to the near face of the ridge beam. In this case, that's 10 ft. Since the pitch is 4-in-12, the rise between those points is 40 in.

Now back to the deck. Measure along the 8-ft. wall, 96¾ in. from the chalked *baseline* of the rake wall. This baseline was snapped out like the other wall lines as a guide for positioning the inside edge of the bottom plate once the wall is framed and raised. But it is also a convenient starting point for the full-scale elevation of the rake wall that you are going to snap out on the deck. In this case, it will be used to represent the bottom of the bottom plate.

The next step is to lay out the

bird's mouth of the rafter above the 96¾-in. mark you just made along the 8-ft. wall. Then take two pieces of 2x scrap and lay them inside the line at approximately a 4-in-12 pitch to represent the rake top plates. The reason you use two pieces of 2x scrap instead of just subtracting 3 in. for the rake top plates is that the vertical thickness of these plates when they are at a pitch will be greater than when they are horizontal. In the case of a 4-in-12, two plates add up to about 3¼ in.

Now make a mark on the chalkline of the 8-ft. wall just below the two scraps. This point represents the top (short point) of the shortest stud in the rake wall, and will be used to establish the line of the rake top plate. To complete this line across the deck, you need to create a large 4-in-12 triangle. In this case, a 4-ft. leg and a 12-ft. leg work out nicely. Actually, this triangle can be any size as long as its proportions are correct, it is close to the length of the rake wall, and it is positioned so that its hypotenuse represents the bottom of the rake top plate.

To lay out this triangle, first make a mark on the rake baseline 12 ft. out from the inside of the 8-ft. wall. From this point, measure up 93½ in. (the height of the 8-ft. wall less the thickness of the two top plates) parallel to the 8-ft. wall, plus another 4 ft. to take care of the rise. Now use a snapline to connect

this point with the one on the 8-ft. wall that you established below the scrap top plates. This is the hypotenuse of the 4-in-12 triangle, and as the bottom edge of the rake top plate it will be used by the framers to cut the tops of the rake studs to length.

To establish the top end of the rake wall, find the point along the rake baseline that represents the near face of the ridge beam. When the rake wall is framed, the top plates will die at the top inside face of this beam. Now snap a line that is parallel to the 8-ft. wall that starts at this point on the rake baseline and ends by intersecting the rake top plate line you just snapped in the center of the deck.

To check your layout, you can use your tape measure to get the height of the longest stud in the rake wall at its long point. Do this by measuring along the chalkline you just established at the inside face of the ridge beam. Stretch your tape from the baseline of the rake wall to the bottom of the top plates, and then subtract 1½ in. for the bottom plate. This measurement should be the same as the shortest stud plus the rise of 40 in. that was figured earlier.

To finish up, determine the position of the ridge beam by laying out the bird's mouth at the top. By subtracting the depth of the ridge beam, you can also determine the length of the posts beneath it.
—*J. P.*

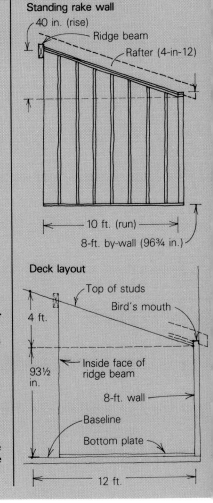

Standing rake wall

40 in. (rise)
Ridge beam
Rafter (4-in-12)
10 ft. (run)
8-ft. by-wall (96¾ in.)

Deck layout

Top of studs
Bird's mouth
4 ft.
Inside face of ridge beam
93½ in.
8-ft. wall
Baseline
Bottom plate
12 ft.

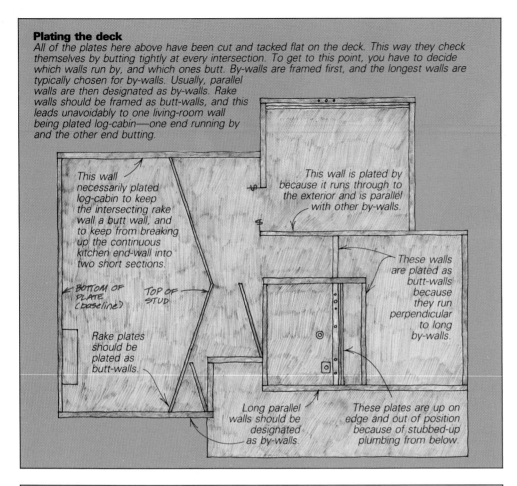

Plating the deck
All of the plates here above have been cut and tacked flat on the deck. This way they check themselves by butting tightly at every intersection. To get to this point, you have to decide which walls run by, and which ones butt. By-walls are framed first, and the longest walls are typically chosen for by-walls. Usually, parallel walls are then designated as by-walls. Rake walls should be framed as butt-walls, and this leads unavoidably to one living-room wall being plated log-cabin—one end running by and the other end butting.

This wall necessarily plated log-cabin to keep the intersecting rake wall a butt wall, and to keep from breaking up the continuous kitchen end-wall into two short sections.

This wall is plated by because it runs through to the exterior and is parallel with other by-walls.

These walls are plated as butt-walls because they run perpendicular to long by-walls.

BOTTOM OF PLATE (baseline)   TOP OF STUD

Rake plates should be plated as butt-walls.

Long parallel walls should be designated as by-walls.

These plates are up on edge and out of position because of stubbed-up plumbing from below.

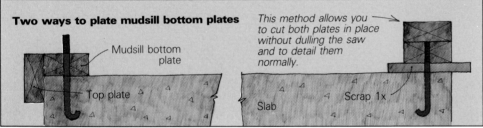

Two ways to plate mudsill bottom plates

Mudsill bottom plate

Top plate

This method allows you to cut both plates in place without dulling the saw and to detail them normally.

Slab   Scrap 1x

Plating. **Peake cuts in the top and bottom plate of a small raceway (left), and tacks each of the two sets of plates together with 8d nails (right) so they can be detailed. One of the first decisions in plating is which walls will run by, and which will butt. Here the long interior wall is a by-wall, and the parallel wall of the raceway is plated the same way.**

let the framers know with directions in keel whether your lines represent the top and bottom of studs, or the plates themselves.

Once the exterior walls have been chalked out, snap the interior walls by paralleling them (measuring out the same distance from several points and connecting them). When a wall ends without butting another wall, indicate where with a symbol that looks like a dollar sign with a single vertical bar. Follow the written dimensions given on the blueprints. Before you use an architect's scale to get a missing dimension, make sure that you can't find it by adding and subtracting others. If a mistake is made—whether it's yours or the draftsman's—you're the one who's going to fix it. Yet it's always better if what you've done reflects the approved plans.

Occasionally, though, you will have to adjust room sizes to accommodate some unanticipated condition. The rule of thumb here is to adjust in the largest rooms. The sizes of smaller spaces are usually dictated by the building code or some prefabricated item.

Plating—Plating involves cutting a top and bottom plate for each wall, tacking them together and laying them in place on the deck. To do this, you must decide at each intersection of walls which ones run through (*by-walls*) and which ones stop short (*butt-walls*). Walls that should be framed first—usually the longest exterior walls—are designated by-walls. Usually, the walls parallel to these will also be plated as by-walls. The framing and plating is simpler if you plate rake walls as butt-walls since the top plates are detailed in place (on their full-scale layout lines) near the middle of the deck. If plated as a by-wall, the top rake plate would lap over the plates of exterior walls that run perpendicular.

What you want to try to avoid in plating is *log-cabining*—building walls that run by at one corner and butt at the other. Such walls will probably have to be slid into position after they're framed and raised, which isn't easy, especially with a heavy wall.

The drawing top left shows the plating for this floor plan, and lists reasons why some walls have been designated butt-walls and others by-walls. Notice that the living-room wall at the top of the drawing has been plated log-cabin. So much for hard-and-fast rules. In this case, the rake wall at the end of the living room had to be a butt wall, and the kitchen wall on the other end had to run by so it wouldn't end up in two extremely short sections. The result is necessary log-cabining.

To do the plating, spread the top and bottom plate stock near the snapped lines. Use long, straight pieces. A crooked top plate can drive a framer crazy when it's time to straighten the raised wall with braces. To make the length on long walls, top plates will have to butt together at the center of a stud. The middle of a solid header is an even better spot. Breaks in the top plate and the double plate (the framer will be supplying this permanent tie between walls) have to be at least 4 ft. apart. That means to stay at least 4 ft. away

Peake pries apart the top and bottom plates on a small section of exterior wall that has to be framed separately because of stubbed-up plumbing. The layout carpenter must constantly be thinking ahead to the framing stages. How the house is plated, and which edges of the plates (inside or outside) the detailing is marked on will have a lot to do with how easily and quickly the framing will go.

from intersecting walls when laying out a break in the top plate, since the double plate of this wall will have to end there.

Very long walls will have to be framed in sections. With an average number of headers in an 8-ft. high 2x4 wall, each carpenter on the site should be able to handle at least 10 lineal feet of wall when it comes time to raise it. If you're going to break a long wall into separate sections, you'll need to end the top plate at the center of a stud. The bottom plate should be broken at the same place.

The plate stock for the bottom plate should be tacked flat to a wood deck along the snapped line with an 8d nail near each end. It's fine if it laps over nearby wall lines, because the next step is to crosscut it in place by eyeballing the chalkline of the intersecting wall. Now lay down the top plate in the same way, cut it and tack it to the bottom plate with two more 8d nails (photos facing page). The only exception to nailing the plates together is a rake wall where the top plate will be left on the deck on its angled layout line.

In the case of a slab, the bottom plate will be a mudsill. I use a bolt marker like Don Dunkley's (text and drawing, p. 19). I usually make mine out of a piece of 1x2 with a joist-hanger nail at 3½ in. and at 5½ in. Once the mudsill is drilled out and set on the bolts, you have to deal with the top plate, which won't tack down to the mudsill because of the bolts.

Some carpenters hang the top plate off the edge of the mudsill (drawing facing page, center left), and then detail the layout across the edge of the top plate and the flat of the mudsill. But the framers will like it better if you shim up the mudsill with 1x scraps (drawing facing page, center right). This allows you to cut both the mudsill and the top plate in place without dulling your sawblade on the concrete, and to detail them normally.

Detailing—After all the plating is complete, detailing can begin. It is done in three stages: recording the information that you've added to the blueprints on the plates, marking out the precise measurements for headers, corners and intersecting walls on the face and edges of the plates, and then adding the stud layout. The drawing on the next page shows the floor plan with the plates fully detailed.

Layout style varies widely from region to region. One difference is in detailing shorthand. Layouts often contain more detailing than is really necessary. For instance, if you indicate where the end of a header falls with a line, and then make an X for the king stud beyond the line, it will be evident to the framer that the trimmer goes on the other side of the line. Writing T for trimmer (or C for cripple, de-

pending on the terminology you use) takes time, and doesn't add any information.

Another difference is the orientation of the plates. Some production carpenters tack the plates together and then toenail them along the chalkline with their edges up, rather than flat. But there are several advantages to running them flat. The first is that they check themselves. They can't be too long or too short because they are laid in the precise positions that they will occupy once they have been framed. Second, the location of headers and wall intersections is easy to see when it's detailed on the top of a flat plate, and won't get overlooked or misframed. Last, all the information necessary for the framers to cut the double plates that interconnect the walls is

marked on the surface to which they will be nailed—the top face of the top plate.

Almost all the marks that you'll put on the plates will be on the top of the top plate, and on one set of edges. The way to determine which set of edges is to approach a pair of tacked-down plates as if you are going to frame the wall. The trick is to figure out from which direction the top plate will be separated, the studs added and the wall raised. With an exterior wall this is easy. The top plate will be walked to the interior of the deck with its top markings facing the opposite side and its stud layout (which is on the edge) up. This way, once the vertical members have been filled in, the wall can merely be tilted up into place without having to reorient it. Exteri-

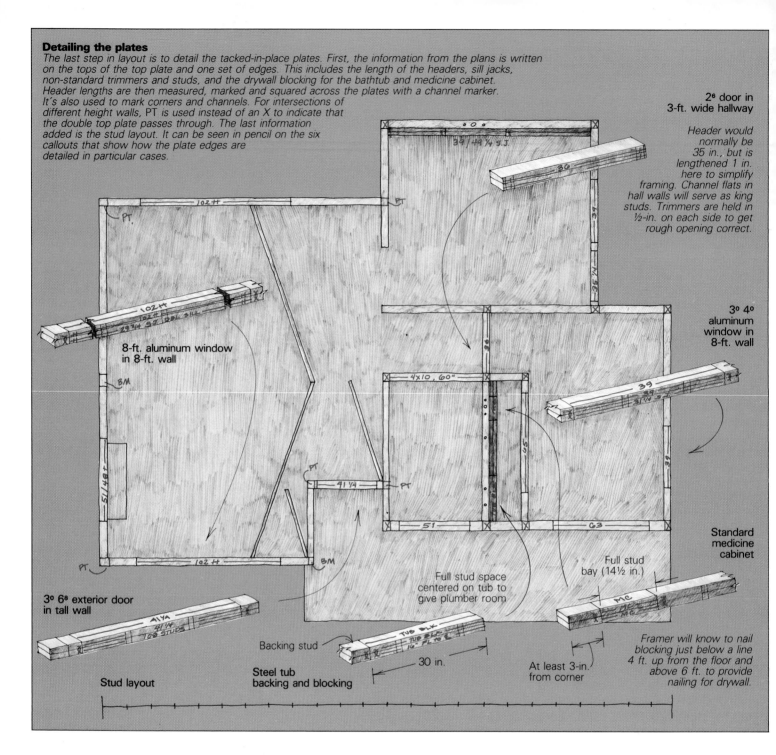

Detailing the plates

The last step in layout is to detail the tacked-in-place plates. First, the information from the plans is written on the tops of the top plate and one set of edges. This includes the length of the headers, sill jacks, non-standard trimmers and studs, and the drywall blocking for the bathtub and medicine cabinet. Header lengths are then measured, marked and squared across the plates with a channel marker. It's also used to mark corners and channels. For intersections of different height walls, PT is used instead of an X to indicate that the double top plate passes through. The last information added is the stud layout. It can be seen in pencil on the six callouts that show how the plate edges are detailed in particular cases.

2⁶ door in 3-ft. wide hallway

Header would normally be 35 in., but is lengthened 1 in. here to simplify framing. Channel flats in hall walls will serve as king studs. Trimmers are held in ½-in. on each side to get rough opening correct.

3⁰ 4⁰ aluminum window in 8-ft. wall

8-ft. aluminum window in 8-ft. wall

Standard medicine cabinet

3⁰ 6⁸ exterior door in tall wall

Full stud space centered on tub to give plumber room

Full stud bay (14½ in.)

Framer will know to nail blocking just below a line 4 ft. up from the floor and above 6 ft. to provide nailing for drywall.

Stud layout

Backing stud

Steel tub backing and blocking

30 in.

At least 3-in. from corner

---

or plates, then, are typically detailed on their outside edges. Interior partitions, since they can often be tilted up from either direction, require the layout carpenter to guess how and where the framers will set up.

You're now ready to do the detailing. Forget your tape measure for the moment, take the plans in hand and walk around the deck copying the information you've written there onto the plates. For each window and door, write its header length on the top of the top plate near where the header will be nailed, and *tt* if it requires double trimmers. Now lean over the plates and mark the outside edge of the top plate with the same information.

As long as you detail the rough-opening dimensions of the windows on the plates when you lay out, you don't have to figure the height of the rough sill or the length of the sill jacks. The framer can do this by measuring

the rough-opening height down from the header, installing the rough sill, and then filling in with the sill jacks. However, if you figure the length of the sill jacks at the layout stage, they can be precut for the framing. Write this length in keel on the outside edge of the bottom plate and follow it with the letters *SJ.* For any walls with studs longer than the standard 92¼ in., write their lengths on the face of the top plate and on the outside edge of one of the plates. Note the location of prefab items, like medicine cabinets *(MC),* and drywall blocking around fixtures like bathtubs *(TUB BLK).* Label rake walls *RAKE.*

At this point you can begin to mark corners and channels on the top and edges of the plates. Both of these are just wall intersections—one at the end of a wall, one in the middle—and are drawn the same width as the plate stock. When these intersections are

framed, they'll need backing studs for wall finish, but since corners and channels are usually made up as units that include their backing studs, you need only to show the intersection of the plates for the framer to get it right.

To detail the corners, use a pencil to scribe a line onto the inside edge of the by-wall plates along the side of the butt-wall plates. Then use a channel marker (see the box on the facing page) to continue that line across the face of the top plate and down the other edges. These lines will show exactly where to nail the walls together once they are framed and raised. Channels aren't much different from corners, except you have two lines to scribe—one on each side of the intersecting plates. Put a big keel *X* between the lines on the face of the top plate, and on each of the outside edges of the plates to show where the double plates of the intersecting wall will create a half-lap. If the walls connect at different heights, the double plates of the by-wall shouldn't be broken for the double plate of the butt wall. The framer should be warned by marking the corner with the letters *PT,* which tell him to *plate through.*

A common framing mistake, usually discovered after the wall is raised, is putting the flat stud of the channel on the wrong side of the wall from the intersecting partition. As long as the framer knows the flat-plating method used here and doesn't reverse the top and bottom plates, where to locate the flat will be obvious. The key is the *X* that marks the channel. Because you are prevented from making any marks on the inside edges of the by-wall plate when you scribe the butt-wall plate to it by the plates themselves, the *X* will get marked only on the opposite set of edges from where the intersection will actually happen. So the framer should nail the channel flush with the edge of the plate that doesn't have an *X.*

Now detail the window and door openings. Following the blueprints, measure accurately to each end of the headers and use your channel marker to square the lines across the top plate and down the outside edges. Make an *X* on the outside of each of these lines to indicate the king stud.

When making an *X* over the edges of the plates, you can save yourself an extra motion by making two intersecting half-circles. This will leave an *X* on each plate when they are separated. The only time I show the location of trimmers is when they are doubled, which I indicate with *tt* (see the 8° 4° living-room window callout in the drawing, previous page).

Interior doors are often placed near the corner of the room they serve. The standard way to frame them is to let the king stud act as one of the backing studs in the channel, compressing space. Once the drywall is hung, this leaves a little less than 3 in. for casing. If the casing is wider than this, it will have to be scribed to the wall. If the space is even narrower and the door is in a butt-wall such as the 2-ft. 6-in. door in the 3-ft. hallway in the plan, you'll have to use the channel flat as a king stud. A useful rule of thumb is that the space left for the trim is about the same as the distance from the studs of the intersecting wall to the inside face of the trimmer.

There are a few special items in bathrooms that need detailing. The medicine cabinet fits between studs, but it will need blocking above at 6 ft. off the floor and below at 4 ft. If the medicine cabinet is near a corner, double the end stud that nails to the channel to give the necessary room for the swing of the cabinet door. Bathtubs and showers should be blocked along their top edges. Detail this by specifying a height on the plate that runs from the floor to the centerline of the blocking. A double stud or flat stud should be laid out to pick up the side and end of the tub or shower. In the drawing, you can see that I've also centered a standard stud space on the plumbing end of the tub to make it easier for the plumber to run supply lines and a drain.

Beam pockets can be detailed with your channel marker, but label them *BM* so they are not confused with channels. The posts under these beams are detailed with their actual width marked on the edge of the plates with keel, and their nominal size written in between these lines. Also give a length for the post if it is different from stud height.

Stud layout—The regular stud layout comes last and is done on the outside edges of the plates. Pull all the outside and inside walls that run perpendicular to the joists and rafters from the same end of the building. Do not break your layout and start again at partitions, but continue the full length of the wall. The standard 16-in. and 24-in. centers are meant to work modularly with 4-ft., 8-ft. and 12-ft. sheet materials. Remember that 16 in. o. c. means from the end of the building to the center of the first stud, so reduce your layout by ¾ in. each time when pulling from the corner (15¼ in. to the first stud, 31¼ in. to the second stud, and so on). This way the plywood sheathing and subfloor will work out with a minimum of cutting and waste.

It's a common mistake to have drywall on your mind when laying out studs. Drywall is relatively cheap, easy to cut, and can be bought in 12-ft. lengths. It should be at the bottom of your list of worries when laying out.

Stud layout should be done in pencil. If you are using a layout stick, you can put your tape measure back in your nail bag once you get started. Scribe along both sides of each finger of the layout stick to mark for the studs, then reposition it farther down the plate and repeat. If you aren't using a layout stick, stretch your tape the length of the wall and make marks at 15¼, 31¼, etc. Then come back with a combination square set at a depth of 3 in., square these marks down the outside edges of the plates, and make an *X* on the leading side of the line. Don't bother to draw a line for both sides of the stud, but don't lose your concentration either when making the Xs. Putting them on the wrong side of the line will cause big headaches later.

Rake top plates can be laid out by stretching a tape from the by-walls they butt, but you'll have to hold the tape perpendicular to the by-wall and keep moving the end of it farther down the by-wall so that the stud centers on the tape will intersect the angled rake plate. A better method is to measure the distance between studs along the angled plate after marking the first few, and then use this increment to mark the studs thereafter.

The last thing to do before you leave the deck is to look over the tops of the plates and make sure that every room has a door in it (a common but embarrassing mistake), and that all the channels are marked. These marks are easy to spot because they are on the top of the plates. Finally, to get the framing off to a good start you can cut all the headers, sills, sill jacks, and specials, as you already have a cut list on the marked-up plans. □

Jud Peake is a contractor and a member of Carpenters Local 36 in Oakland, Calif.

Tools of the trade

Most production layout tools were born of necessity on the site and made with available materials on a rainy day. One such tool is the channel marker (middle photo), a simple square made out of short pieces of plate stock and used for outlining corners and channels. It should have a leg 3 in. long (the depth of two 2x plates) and another leg 3½ in. long (the width of a 2x4). Both legs are 3½ in. wide. I make a more durable version with aluminum flat stock that includes a 1½-in. flange at the top. By turning the square over, you can lay out the thickness of a stud with this flange.

Two more tools that will speed things up are a layout stick (top photo) and a keel/pencil holder (bottom photo). This last item is just a short piece of ½-in. clear plastic tubing that will take a carpenter's pencil in one end, and your keel in the other. Layout sticks can be made out of standard aluminum extrusions riveted together. The 1½-in. wide and 3-in. long fingers on mine are laid out for 16-in. centers and 24-in. centers. I even threw a hinge into my stick so that it could fold up to fit in a standard carpenter's toolbox. —*J. P.*

Parallel-Chord Floor Trusses

Strong and efficient, floor trusses may someday replace wood joists as the builder's favorite floor frame

by E. Kurt Albaugh

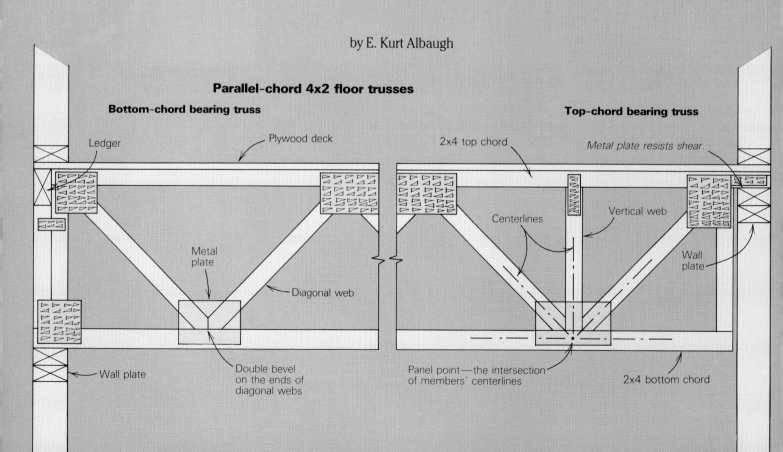

Parallel-chord 4x2 floor trusses

Bottom-chord bearing truss

Top-chord bearing truss

Ledger

Plywood deck

2x4 top chord

Metal plate resists shear.

Metal plate

Centerlines

Vertical web

Diagonal web

Wall plate

Wall plate

Double bevel on the ends of diagonal webs

Panel point—the intersection of members' centerlines

2x4 bottom chord

Once associated more with commercial construction, structural trusses are becoming increasingly popular among home builders. Though most builders are familiar with roof trusses, fewer builders realize that floor trusses can be used quite effectively in residential construction, too. They offer a number of structural and economic advantages, and can easily be incorporated into the design of a home without significantly changing the way a builder builds.

Structural advantages of trusses—Traditionally, the size of rooms in a home has been based largely on the span limitations of standard wood joists. Floor trusses with the same depth as joists can be used over longer spans, and this means that rooms can be larger, with less space obstructed by columns or unnecessary partitions. Trusses have a much greater variety of depths than wood joists do, and therefore a much wider range of spans and strength.

Trusses rarely warp. With a joist floor, natural warpage in the members can lead to an uneven floor deck. This problem can be reduced somewhat by culling out the warped joists before installation, but that increases wood waste.

Another advantage of floor trusses is that they permit ducting, plumbing and electrical service to be run easily between the open webs. With a joist floor system, holes must often be cut through each joist in order to run electrical and plumbing lines.

Since floor trusses are sized by the fabricator, they're delivered to the job site in lengths that meet the specific requirements of the project. With joists, the material must often be cut on site to fit, which wastes wood and increases the handling of material. Installation of a truss is relatively easy, because the truss edge is 3½ in. wide. This makes it more stable while it rests on the wall plates, and easier to nail to.

A floor-truss system is engineered, while a joist system usually isn't. This means that greater floor loads can usually be carried, and deflections are better controlled. Engineering also makes for a more efficient use of material (strength vs. weight), resulting in a lower per-square-foot cost.

Cost advantages of trusses—The builders I've worked with who use floor trusses believe that a floor-truss system is cost effective. But exactly how much so is difficult to determine since labor and material costs vary greatly throughout the United States. As a general rule for the Houston area, material costs are about $.90 per sq. ft. of floor (with trusses 24 in. o. c.). If the job is a long way from the manufacturer, shipping can increase the costs significantly. A joist floor system, on the other hand, costs about $.85 per sq. ft. of floor. This means that material costs for a truss floor are about 6% higher than for a joist floor system. But when the time saved for installing a truss floor (as much as 40%) is taken into account, the truss floor turns out to be about 10% less expensive overall.

What are the time savings? Quicker framing means a faster construction schedule, with attendant savings on construction financing. And because plumbers can route the pipes through the open webs of the trusses, no time is wasted by having to bore holes through joists. The electrician and HVAC (heating, ventilating and air conditioning) installers save time as well. Exactly how much time each tradesperson saves is hard to determine, but it can add up over the course of a job.

On the production side, most wood-truss fabricators find that there is little difference in production costs between 12-in., 14-in. and 16-in. deep trusses, so these tend to be comparable in cost to the builder. But when trusses get deeper than 16 in., material costs increase significantly because web material other than #3 shorts must be bought (now you know where mill ends go).

After pricing a lot of wood-chord trusses here in the Houston area, I quickly found out that the size of a builder's truss order is the greatest single factor affecting the cost of trusses. If there are several truss fabricators in your area, each one may have a different business approach. One supplier will probably favor small-quantity, custom truss orders while another will specialize in mass production and large orders. Shopping around for the best price often means shopping for the fabricator whose business most fits yours.

Truss anatomy—A floor truss has only three components—chords, webs and connector plates. Each one is critical to the function of the truss. The wood chords, or outer members, are held rigidly apart by wood or metal webs. The strength-to-weight ratio of floor trusses is higher than that of solid-wood joists because the structural configuration of the truss converts the bending moments and shear forces (produced by loading) into compressive and tensile forces. These forces are directed through the individual truss members and transferred to walls.

Chords. The type of floor truss used most frequently in residential floor systems is called the parallel-chord 4x2 truss. As shown in the drawings on the facing page, the chords are evenly spaced from each other, and the designation "4x2" identifies them as 2x4 lumber with the wide surfaces facing each other. This configuration increases the structural efficiency of shallow trusses, and provides them with a larger bearing area on wall plates. It also gives trusses additional lateral rigidity to resist damage during transport and installation.

In residential construction, a floor truss will most often bear on the underside of its bottom chord (left drawing, facing page), just as a joist bears on its bottom edge. But because of the structural versatility of a truss, it can also be designed to bear on the underside of its top chord. In this case, the bottom chord is shortened and the truss hangs between walls, instead of resting on top of them (see right drawing, facing page). This can be a useful feature when the overall height of a building is restricted. But since height isn't usually a problem in residential construction, top-chord bearing trusses are seen more frequently in commercial work.

Chords can be made of spliced lumber as long as a metal connector plate is used to join the pieces. Chords are kiln-dried, and in the southern United States, they're generally made from No. 1 KD southern yellow pine. Truss fabri-
cators commonly use either machine stress-rated lumber or visually graded lumber for truss chords, depending on the cost and availability of the lumber grade. During its early years, the wood-truss industry used only visually graded lumber, and trusses were designed conservatively to compensate for substantial variations in the strength and stiffness properties of the wood. Some time ago, engineers at Purdue University developed a technique to test and grade lumber by machine. Lumber graded by this process is called machine stress-rated (MSR) lumber, and truss fabricators are using it with increasing frequency for chord stock. Because every piece of MSR lumber is mechanically tested for stiffness and given a categorical strength rating, trusses can be designed to maximize the use of the wood's strength.

Webs. The members that connect the chords are called webs. Diagonal webs primarily resist the shear forces in the truss, and they are usually positioned at 45° to the chords. Vertical webs, which are placed perpendicular to the chords, are used at critical load-transfer points where additional strength is required. They also are used to reduce the loads going through the diagonal members.

Since the strength of the webs is not as critical as is the strength of the chords, a lower-quality wood can be used, such as #3 KD southern pine. Wood is the material used most frequently for webs, but metal webs are also used. Metal webs are stamped from 16-ga., 18-ga. or 20-ga. galvanized steel. Several truss manufacturers have designed the metal web to incorporate a connecting plate, thus reducing the number of pieces required to assemble the truss, as shown in the drawing below. Metal-web trusses provide greater clear spans for any given truss depth than wood-web trusses. The opening between the webs is larger, too, which allows more room for HVAC ducting. And they're lighter, so they are easier to install. But a major disadvantage of metal-web trusses is that a wide range of truss depths isn't always available. Also, metal webs are more susceptible to damage during transport and installation.

In a truss with wood webs, the ends of the diagonal webs are double-beveled wherever they meet a vertical web and the chord. There is an important structural reason for this cut. The objective in the design of any truss is for the centerlines of the chord and adjoining pairs of webs to intersect at one point. This is called the panel point. When centerlines don't intersect at

Types of metal webs

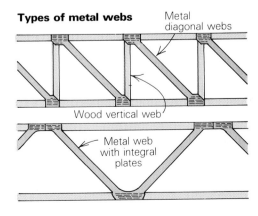

Metal diagonal webs

Wood vertical web

Metal web with integral plates

Floor trusses can be cantilevered over longer distances than wood joists. These trusses will support a balcony. Pressure-treated wood was used for the end web as a precaution against rot caused by possible water infiltration. The double vertical web member at the wall provides additional support for the deck loads on the cantilever.

a panel point, additional and undesirable stresses are introduced at the joint.

There are several different types of parallel-chord floor trusses, and what makes each one different is the arrangement of its webs. Depending on their arrangement, the webs will be either in compression or tension, and this dictates how loads are transferred from the truss to the structure supporting it. Pratt, Howe and Warren trusses were named after their respective developers. Howe and Pratt designed their trusses for railroad bridges. Two giant timber-framed trusses could span a river or a canyon, and trains would run over tracks that stretched alongside the bottom chords.

Connector plates. In 1952, A. Carol Sanford invented the toothed metal connector plate, which eliminated the need for nailing and gluing truss plates. Metal connector plates substantially reduced the cost of trusses by allowing them to be mass produced.

Variations of the toothed connector plate are now commonly used to assemble trusses. They are made from 16-ga., 18-ga. or 20-ga. hot-dipped galvanized steel, which is punched to form numerous metal prongs, or "teeth," that extend outward from one side of the plate. When embedded in the wood of the truss (usually by a hydraulic press), these teeth give the plate its holding power. Specifications for designing trusses using metal connector plates are available from the Truss Plate Institute, Inc. (583 D'Onofrio Drive, Suite 200, Madison, Wis. 53719), which also distributes test and research data.

If you've had a chance to examine connector plates, you may have noticed that they are not always the same size. As they increase in size, they provide more embedded teeth, and therefore more holding power. Larger plates are used in joints with higher stresses.

Deflection—The design of a truss is usually governed more by bending limitations than by anything else. Too much bending, or *deflection*, can make a floor noticeably springy and result in cracks in the finished ceiling below. The

American Institute of Timber Construction (333 W. Hampden Ave., Suite 712, Englewood, Colo. 80110) recommends deflection limitations for trusses of L/360 for the live-load portion of the total load and L/240 for the total load, where L is the span in inches. Live loads are calculated only if they are expected to be unusual—for example, placing a pair of pianos on the second floor. Generally, though, the total load calculation is used. Total load includes the weight of everything attached to or bearing on the top and bottom chords. It includes the dead weight of floor and ceiling materials, as well as the live loads of people, pianos and furniture.

To see how deflection limitations are used, suppose a floor truss must be selected to span 22 ft. between two walls. The maximum allowable total load deflection is calculated by converting the span distance to inches (22 x 12 = 264) and dividing this by 240. So a truss must be selected that will deflect no more than 1.1 inch at mid-span. In other words, if the truss is subjected to all the loads anticipated, the most it may deflect downward is 1.1 in. Remember that this is total deflection, including that caused by dead weight; it doesn't mean that every footstep will cause the truss to deflect 1.1 in.

Deflection limitations provide design parameters for the engineer in selecting an appropriate truss. Knowing the maximum deflection allowable over a particular span, the engineer can determine a truss depth that will deflect less than the maximum allowable distance. This can be done by mathematical calculation or by consulting data from truss-plate manufacturers concerning their products.

A floor system consists of the deck material and supporting network of trusses, and both components function together to transfer floor loads to the walls and foundation. But when trusses are engineered for a particular application, the effect of the deck material on truss stiffness is not taken into account. The floor-system strength is based solely on the stiffness of the truss itself. Because of this, the deflection calculations are on the conservative side.

Choosing a truss—Figuring the total loading on a floor system involves some calculation, but you don't necessarily have to hire an engineer or an architect to size trusses for your project. Engineering is provided either by the truss fabricator or by the plate manufacturer whose plates are purchased by the truss fabricator. These people work together to ensure that the combinations of plates, webs and chords are structurally sound.

To size a truss, the fabricator will first review your plans and look for any unusual circumstances that could affect the truss design. If nothing unusual is found, the truss design and depth will be chosen from reference manuals supplied by the plate manufacturer. These manuals list the standard truss designs and include span tables for each. The span tables simplify all the engineering variables into a very usable form, and are based on design formulas that come from actual laboratory tests of trusses. Spans are listed for four commonly used truss spacings: 12 in, 16 in., 19.2 in. and 24 in. The most common spacing for residential construction is 24 in., but trusses can be placed at any spacing that will allow the floor system to carry the loads specified in the building codes. All the spans listed in the tables assume the deflection limitations of L/360 or L/240.

Since truss-plate size is a critical factor in truss design, the plate manufacturer will often supply pre-certified engineering for the trusses, stating the limitations with in which they will meet certain performance standards, such as the deflection limitations.

If a standard design can be found to meet your needs, the truss fabricator will provide you with a certified drawing of it. Proof of this certification is the engineer's stamp on each page of the drawing.

If the design of your structure is unusual, with particularly long spans, heavy loads or unusual support conditions, it may call for a floor truss that has to be engineered specifically for your project. In this case you'll need to arrange for an outside engineer to do the calculations.

Typically, parallel-chord floor trusses with wood webs are available in depths ranging from 12 in. to 24 in., in 1-in. increments. The most common depths are 14 in. and 16 in. One plate manufacturer has designed a metal-web truss with the same actual depth dimensions as 2x8, 2x10 and 2x12 solid-wood joists. These smaller sizes are interchangeable with an ordinary joist-floor system.

The amount of space between webs is another thing to consider when you choose a truss. When ducts will be routed under the floor, the depth of the truss may be dictated by the size of the ducts. Usually this isn't a problem, since most fabricators have a standard truss design that includes a chase opening. To create a chase opening, one web in mid-span is removed to provide space for large ducts. The truss doesn't collapse on account of the missing web because shear forces are minimal at mid-span.

One last note about choosing trusses. Before you leave the fabricator's office, make sure that your order is complete. If you run short of joists, you can dash off to the lumberyard. But since trusses are precisely fabricated to your specifications, you will probably have to re-order the ones that you forgot. This can be expensive, and time-consuming as well.

Truss bearing details—A critical aspect of floor-truss design and installation is the amount and location of bearing surfaces, since the accumulated loads of the truss are concentrated here. Bearing details vary, depending on how and where the truss is being used.

For the simple-beam condition (drawing A, right), the truss is supported at each end, and rests on its lower chord, in joist fashion. It can also rest on its upper chord, and will carry the same amount of weight as an otherwise identical bottom-chord bearing truss.

A floor truss designed as a continuous beam will be supported on each end, as well as at one or more points in between. To ensure proper load transfer at the intermediate support points (a column or partition, perhaps), a vertical web member should be located directly over the bearing area (drawing B). Some truss fabricators will identify the intermediate load-bearing points on a continuous-span truss by stapling a green or red warning note to the area. These warnings are particularly helpful when you install the truss, since they help position the truss and keep you from accidentally installing it upside down.

When floor trusses are to be cantilevered (drawing C), to support a balcony, for example, it is particularly important that provisions be made for extra support at the bearing points (photo, facing page). The truss should always be designed so that a panel point rests over the bearing point. This transfers the load from the truss directly to the supporting element. To strengthen the area even more, the bottom and top chords can be doubled near the support point of the cantilever.

Trusses can also be used in multiples. In drawing D, two of them meet over an interior partition, while in drawing E, a steel beam provides the intermediate support.

Strengthening trusses—When floor systems are designed, the usual approach is to specify a single truss depth that will work for the entire system, regardless of the various span distances. This simplifies material orders and eliminates the possibility of mix-ups at the job site. Situations do arise, however, when it isn't practical or cost-effective to increase the depth of all the trusses in order to beef up just a few, and in these cases the truss fabricator can increase the stiffness of individual trusses.

Chords can be strengthened across the entire length of the truss by reinforcing the top and bottom chords with a second layer of wood, fastened into place with metal plates, nails or glue. Side-by-side floor trusses are often used to carry greater floor loads (bottom photo, next page), in a technique akin to sistering wood joists. Other options include using larger truss plates, stronger wood for the chords, or doubled webs in critical shear areas of the truss.

Sometimes floor trusses have a slight lengthwise curve that's built into them by the manufacturer. This curve is called *camber*, and it helps

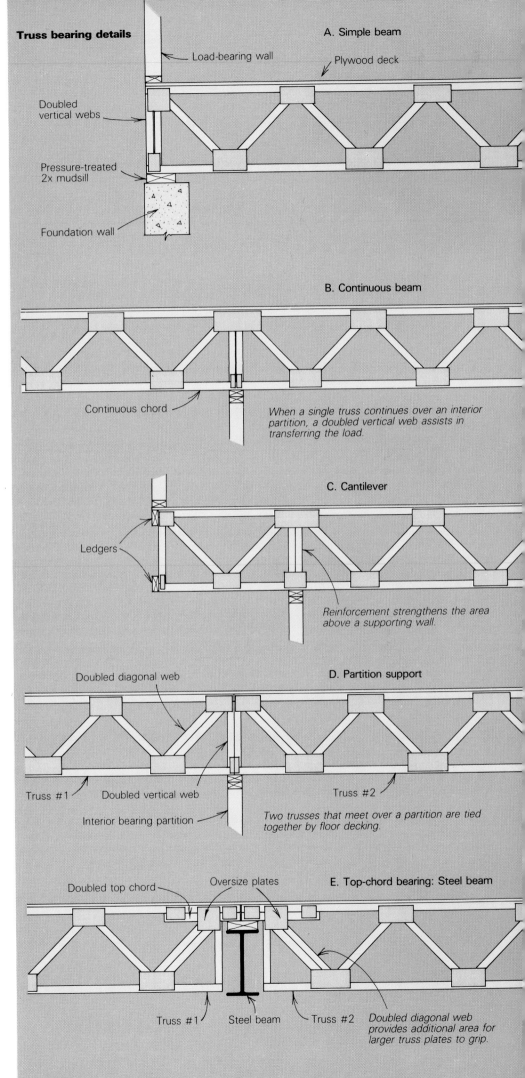

Truss bearing details

A. Simple beam

Load-bearing wall
Plywood deck
Doubled vertical webs
Pressure-treated 2x mudsill
Foundation wall

B. Continuous beam

Continuous chord

When a single truss continues over an interior partition, a doubled vertical web assists in transferring the load.

C. Cantilever

Ledgers

Reinforcement strengthens the area above a supporting wall.

D. Partition support

Doubled diagonal web
Truss #1
Doubled vertical web
Interior bearing partition
Truss #2

Two trusses that meet over a partition are tied together by floor decking.

E. Top-chord bearing: Steel beam

Doubled top chord
Oversize plates
Truss #1
Steel beam
Truss #2

Doubled diagonal web provides additional area for larger truss plates to grip.

Before the decking is nailed down, trusses should be measured at mid-span to make sure they haven't swayed out of alignment (note the tape measure in the photo). A pair of 2x4s can then be nailed to the trusses at the chase opening to ensure that the top and bottom chords stay at the proper spacing. These 2x4s are not strongbacks, however; strongbacks are usually 2x10s or 2x12s.

After the trusses are toenailed to the top plate, a ledger is nailed in place along their top corners. This further stabilizes the floor and provides an additional horizontal nailing surface for siding. To support additional loads, particularly at the edges of floor openings, trusses can be doubled.

too. They should be closely fitted. Truss-plate teeth should be evenly embedded into the wood for proper load transfer.

Installing floor trusses—A certain amount of care must be taken when installing floor trusses, but otherwise the installation isn't much different from the installation of wood joists. When trusses are delivered to the job site, be sure to check them against the design drawings. Any structural peculiarities, like bearing blocks or doubled chords, should be verified. A bearing block is the extra vertical web that's put in a truss over any support point. Watch for the identification tags on the truss to help you find the bearing blocks.

Metal-web trusses should be inspected carefully to ensure that no webs have been bent. As a general rule, most metal-web trusses are designed to put the webs in tension, so a bent web is not always a problem. But it's best not to take chances. Wood webs should be checked for looseness or damage, as should the plates.

As a rule, the truss fabricator will deliver the trusses to the job site. Make sure you have enough help on site to unload and place the trusses. Be careful when handling them because the edges of a sharp truss plate or metal web can cut. Wear gloves.

Some builders have told me that the reason they don't like to use trusses is because they have to hire a crane to lift large ones into position. I firmly believe that even large trusses can be manhandled up to first-floor wall plates, though the task is a little more difficult when you get to second-floor plates. If you decide to use a crane for the job, you might as well use it to stack the plywood up there, too. Just be sure to nail the trusses securely into place first, and be sure you don't overload them. Bottom-chord-bearing trusses can be further stabilized by temporarily nailing a stringer board to the ends of the trusses.

As the trusses are being set into place, make sure they're right side up. They're sometimes installed upside down by accident, and when this happens, webs and chords that were designed for compression will be in tension. Truss failure is the likely result of this mistake.

Once the trusses have been toenailed to the wall plate (if need be, you can nail through one of the holes in the connector plate) a ledger board (photo below left) is nailed to their top edge to stabilize them and provide a continous horizontal nailing surface for siding. The ledger fits into a notch created for this purpose by the truss fabricator.

If the top or bottom of the truss system is to be exposed when the structure is complete, it should be permanently tied together somewhere at mid-span. If the top chords are exposed to an unfloored attic, for example, the trusses should be braced with 2x material at least every 3 ft. along their length. If the bottom chords are exposed to an unfinished basement, bracing should not exceed 10-ft. intervals. □

E. Kurt Albaugh, P. E., is a consulting engineer based in Houston, Tex. He has a patent pending for a new type of truss connector plate.

to eliminate the visual effects of deflection and to control cracking. But camber does not add strength—it's simply a matter of appearance.

A common way to strengthen a floor system is to install a strongback, usually a 2x10 or 2x12, between the web openings to help distribute loading to adjacent trusses. Sometimes two 2x4s are slipped on edge through the web openings of a series of trusses at mid-span, and then nailed to a vertical web member, as shown in the top photo. This doesn't add much strength, but it helps to maintain truss alignment.

Choosing a truss fabricator—The strength of trusses is affected by the quality of their construction, so purchase your trusses from a reputable fabricator. If you aren't familiar with fabricators in your area, visit their production shops or look at samples of their work at job sites before you buy.

When you examine webs and chords, look at their alignment. Make sure that the centerlines of all the members intersect at a panel point, and that this point is adequately covered by a truss plate. Take a close look at the wood joints,

Snapping a line. A chalkline can be snapped across the tops of studs and cripples to mark a cutline. Before anything is cut to length, the framers will set the top plate on edge above the line and mark the framing layout on it. *Photo by Ron Turk.*

Building Rake Walls
Two time-saving layout methods

by Larry Haun

Most wall layout is quite simple. The process of transferring dimensions from prints to concrete slab or subfloor usually consists of little more than snapping a series of chalklines to form squares and rectangles. On occasion, however, plans will call for a room with a cathedral ceiling that follows the pitch of the roof. Here rafters double as joists, rising upward from an outside wall to the ridge. Gable-end walls in these rooms are called rake walls, and laying one out isn't much more difficult than laying out a regular wall. But being aware of a couple of simple techniques will speed up the process. The methods I discuss here have served me well for the past 30 years (for another approach to building rake walls, see the article on pp. 42-43).

The location of the bottom 2x plate of a rake wall is laid out on the floor like any other wall. The location of the rake wall's top plate is chalklined out at an angle from the top of the shortest stud. This way, the framer can build the wall without making any further calculations, even though each stud will be a different length. The angle of the top plate is determined by the pitch of the roof.

A calculated solution—There are two fairly easy ways of laying out rake walls. The first calls for a pocket calculator, which is used to determine the difference in length between the shortest and longest studs. The shortest stud is normally a standard length, 92¼ in., so once you've established the *difference* in length between shortest and longest, you know the *actual* length of the longest stud. With the heights of both studs established, you'll know the position of the top plate, as well.

To determine the difference in length between the shortest and longest studs in a rake wall, you need to know both the length of the wall and the pitch of the roof. For example, in a house that's 33 ft. wide, a rake wall running to the center of the roof is 16 ft. 6 in. long. With a 6-in-12 roof pitch, multiply 6 by 16 ft. 6 in. (6 in. of rise for every foot of run and 16 ft. 6 in. of run) for a result of 99 in. Add 99 in. to the length of your shortest stud, and you've got the length of your longest stud—191¼ in.

Now go back to the subfloor to lay out the top plate (drawing next page). First, go to the end of the chalkline marking the bottom plate, where the plan indicates the low point of the rake wall. Usually this will be at an exterior wall, but check the plans for the exact location of the shortest stud. Measure up 92¼ in. on the subfloor and mark that point. Next, come over 16 ft. 6 in. along the same chalk-line to the house's center. Measure up 191¼ in. from there for the long stud and mark that point. Make sure your measurements are perpendicular to the chalkline. Connect the two points with a chalkline, and you've established the location of your top plate. Intermediate studs can now be cut to length without any further calculations.

No math, no sweat—Not every good carpenter tackles problems this way, however, and calculators still haven't become commonplace in most tool belts. Another method of laying out rake walls, developed by framers, dispenses with calculation altogether. The trick is to work on a 12-ft. grid and to figure the pitch in feet rather than inches (drawing facing page).

Let's look at the same problem again: a 6-in-12 pitch and a 16-ft. 6-in. wall. Measure up 92¼ in. from the bottom-plate chalkline at the low end of the rake wall. Mark that point; the height of the short stud hasn't changed. Next, come over 12 ft. along the bottom-plate chalkline and again measure up 92¼ in. perpendicular to the chalkline. Mark this point. So far, all you've done is lay out a rectangle that is 92¼ in. on the short sides and 12 ft. on the long sides.

From the last point, at the 92¼-in. mark, measure straight up in feet whatever the roof pitch is in inches. In this example, because you're working with a 6-in-12 roof pitch, measure up 6 ft. (the rise for a 12-ft. run) and mark that point. Snap a line between this point and the top of your short stud, and you've got your

roof pitch. Because your wall is longer than 12 ft., it's necessary to extend this top-plate chalkline several feet. Complete the layout by snapping a chalkline that will represent the outside edge of the longest stud at 16 ft. 6 in. You now have a full-size layout of the rake wall. This process works regardless of the pitch of the roof or the length of the wall and usually can be completed in just a few minutes.

You may run into situations where there isn't enough floor space to lay out the rake wall using this method. When that occurs, simply cut the wall length in half and lay out the pitch as a 3-in-6 instead of 6-in-12. This way, you only need 6 ft. of floor space.

Remember that the lines snapped on the floor show the length of the shortest and longest studs (at their outside edges), the

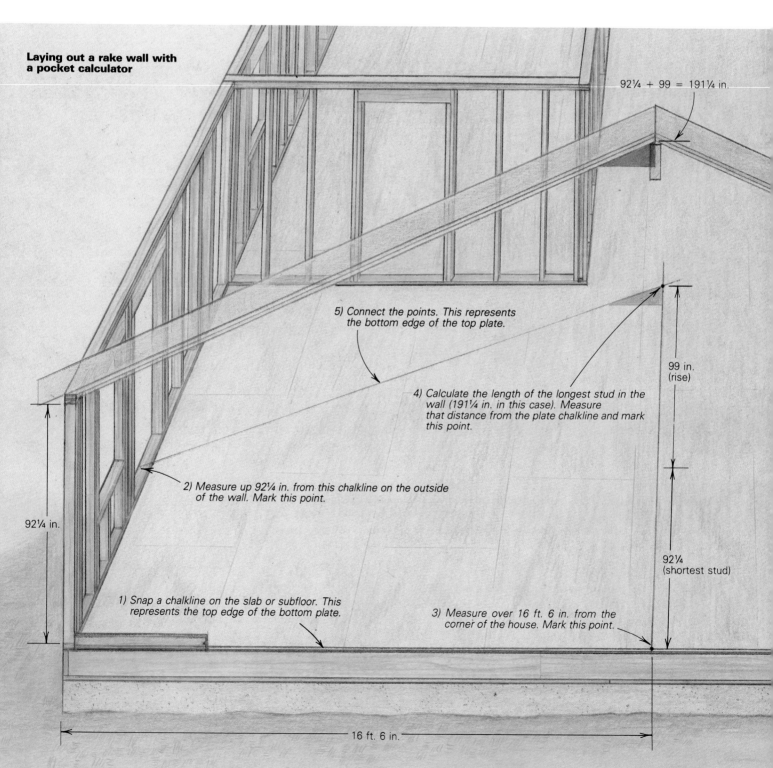

Laying out a rake wall with a pocket calculator

92¼ + 99 = 191¼ in.

5) Connect the points. This represents the bottom edge of the top plate.

4) Calculate the length of the longest stud in the wall (191¼ in. in this case). Measure that distance from the plate chalkline and mark this point.

99 in. (rise)

92¼ (shortest stud)

2) Measure up 92¼ in. from this chalkline on the outside of the wall. Mark this point.

92¼ in.

1) Snap a chalkline on the slab or subfloor. This represents the top edge of the bottom plate.

3) Measure over 16 ft. 6 in. from the corner of the house. Mark this point.

16 ft. 6 in.

Drawings: Bob Goodfellow

roof pitch and the length of the wall. The bottom plate goes below the bottom line, and the top plate goes above the top line as the wall is framed.

Framing a rake wall—Once you've got the perimeter of your rake wall laid out, mark *two* bottom plates (you'll see why in a moment) with stud, window and door locations. Place a stud at every layout mark, letting them extend a little beyond the top-plate chalkline. Nail these studs to one bottom plate, including any trimmers and king studs. Also nail in any headers at this stage. Cripples on top of the headers need to run past the top-plate chalkline, just as the studs do. Next, position the bottom plate so that its top edge is on, but below, the chalkline. Tack it in place with a

few 8d nails to make sure it stays straight, and use the extra bottom plate as a layout guide to align the loose top ends of the studs. Then pull a chalkline on the studs directly over the roof-pitch line and snap it to mark the studs for length (photo, p. 39). Before cutting the studs to length, bring in the piece of lumber that will be your new top plate, place it on edge directly above the chalkline, mark it for length and indicate on it where the studs will be nailed once they are cut.

Now it's time to cut the studs to length. If the saw's shoe tilts in the right direction to make the cut, set it at the proper degree for the roof pitch (26½° for a 6-in-12 pitch). If the angles marked on the studs run opposite the direction in which your saw tilts, first cut them square and then make the second cut

at the correct angle. The next step is to nail on the top and double plates. Lap the double plate over 3½ in. at the low end to tie the two walls together at the corner.

How much precision is necessary?—It's been my experience that carpenters often spend too much time on rake walls, trying to build them to extremely fine tolerances. It usually doesn't matter if these walls get built a little high or a little low. With a site-built roof, I actually like to run the rake wall at least 1 in. high so that a good tie can be made between it and the rafter sitting atop it.   □

Larry Haun lives in Los Angeles, Calif., and is a member of Local 409; he was a longtime teacher in the apprenticeship program.

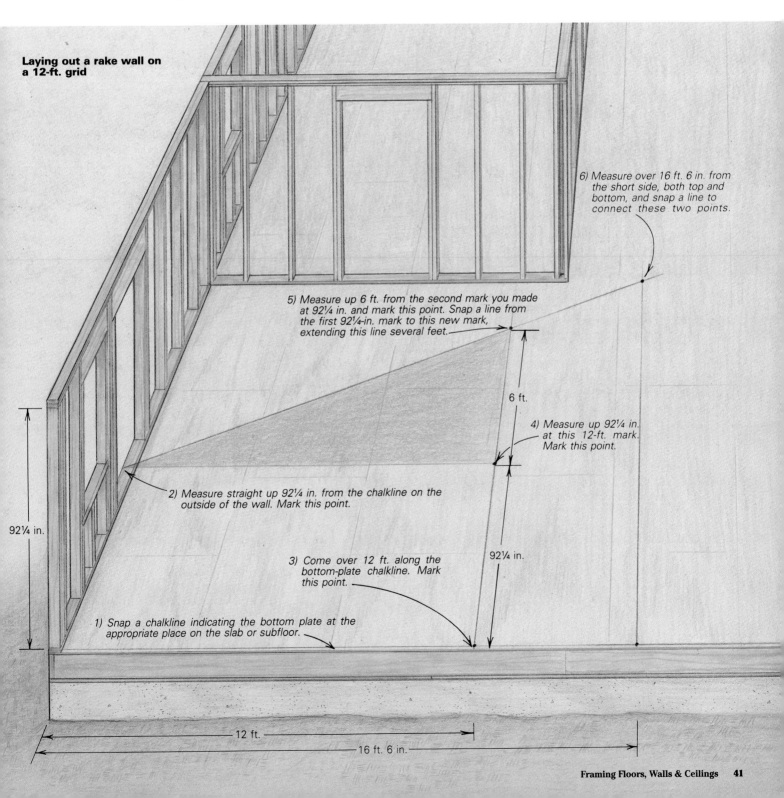

Laying out a rake wall on a 12-ft. grid

6) Measure over 16 ft. 6 in. from the short side, both top and bottom, and snap a line to connect these two points.

5) Measure up 6 ft. from the second mark you made at 92¼ in. and mark this point. Snap a line from the first 92¼-in. mark to this new mark, extending this line several feet.

6 ft.

4) Measure up 92¼ in. at this 12-ft. mark. Mark this point.

2) Measure straight up 92¼ in. from the chalkline on the outside of the wall. Mark this point.

92¼ in.

92¼ in.

3) Come over 12 ft. along the bottom-plate chalkline. Mark this point.

1) Snap a chalkline indicating the bottom plate at the appropriate place on the slab or subfloor.

12 ft.

16 ft. 6 in.

Balloon-Framing a Rake Wall

How one builder stiffens walls by running studs from floor to roofline

by Sean Sheehan

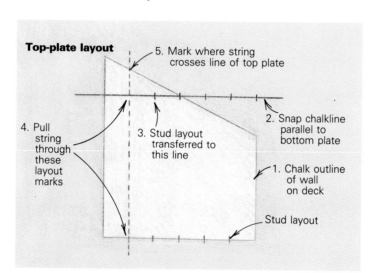

Top-plate layout

5. Mark where string crosses line of top plate

4. Pull string through these layout marks

3. Stud layout transferred to this line

2. Snap chalkline parallel to bottom plate

1. Chalk outline of wall on deck

Stud layout

The concept

Think of the rake wall as a triangle on top of a rectangle.

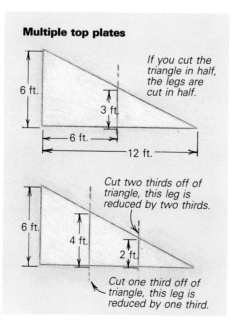

Multiple top plates

6 ft.

3 ft.

6 ft.

12 ft.

If you cut the triangle in half, the legs are cut in half.

6 ft.

4 ft.

2 ft.

Cut two thirds off of triangle, this leg is reduced by two thirds.

Cut one third off of triangle, this leg is reduced by one third.

ere in Montana, the wind is something you can count on. The broad mountain valleys that grace this state are, among other things, nature's wind tunnels (the high plains are called windswept for good reason). A builder must contend with the wind during all phases of a project, and any building should be designed with the wind in mind.

One technique that our crew uses to increase the wind resistance of a rake wall (a wall whose top plate follows the incline of the roof) is balloon-framing. Whenever we can, we extend the rake-wall studs all the way from the floor to the roofline rather than frame a conventional wall and stand a truss on it or fill in above the wall with gable-end studs.

Balloon- vs. platform-framing—In balloon-framing, studs run continuously from foundation to roof. The second floor, if there is one, hangs from the studs. In platform-framing, which evolved from balloon-framing as a safer and more efficient form of construction, the second floor is built on top of the first-floor walls. Then the second-floor walls are framed on top of the platform (hence the name). With this system the top plates of the first-floor walls serve as firestops; in balloon-framing, firestops have to be added. Platform-framing also requires shorter studs, which are easier to handle, and provides a safe platform (the second floor) on which to work, rather than requiring the carpenter to build walls 16 ft. in the air.

Nonetheless, there are times when balloon-framing makes sense. I consider it essential when building a tall, window-filled rake wall in a home with high cathedral ceilings. Even when sheathed with plywood, the platform-framed version of this wall can literally billow in the wind. The top plates that divide a platform-framed wall from the rake-wall studs above it create a break line. When the wind pushes against such an arrangement and there is no interior structure (such as a second floor or a partition wall) to resist it, the wall flexes at the break line.

The structural integrity of balloon-framing can be undermined by careless placement of windows and doors. We try to ensure that studs in the center third of a rake wall are left intact. If this is impossible, we double up king studs, or sometimes triple-stud the center of a wall if there are windows on both sides of center. The object is a stiffer wall, so we leave enough of the studs in one piece to achieve this goal.

We use two basic methods to balloon-frame a rake wall. If we have the space, we build it on the first-floor deck. We usually divide a peaked wall into two wedge-shaped walls and nail them together after the walls are up. This provides more work space on the deck and puts a double stud in the center that runs to the peak. If we don't have room on the deck, then we build the wall in place, or "in the air," as we call it.

First, the math—Usually, the eave height and the length of the wall are known, and I have to determine the peak height, the length of the top plate and the length of each stud. The peak height is determined by the pitch of the roof we're building. Let's assume a 6-in-12 pitch, a wall length of 12 ft. and an eave height of 8 ft. Over 12 ft., a 6-in-12 pitch will rise 6 ft. Add 6 ft. to the height of the eave for a peak height of 14 ft. To find the length of the top plate, it's helpful to think of the wall as a right triangle on top of a rectangle (center left drawing). The rectangle is 12 ft. by 8 ft. (the length of the wall by the height at the eave), and the right triangle is 6 ft. high and 12 ft. long with an unknown hypotenuse (the top plate). I use the Pythagorean theorem ($a^2 + b^2 = c^2$) to find the length of the top plate.

In this case, $a = 6$ (the length of one leg) and $b = 12$ (the length of the other leg). Plugging these numbers into the formula, you'll find that c^2 must be equal to 180. Now all you need is to find the square root of 180, which is 13.416 ft., or 13 ft. 5 in.

On the deck—Once I have the length of the top and bottom plates and the height of the wall at both ends, I chalkline a full-scale pattern of the wall on the deck, being careful to keep the pattern square.

To mark the stud locations on the top plate, another carpenter and I snap a line parallel to

Drawings: Michael Mandarano

the bottom plate that crosses the top plate somewhere near the middle (top drawing facing page). Next we transfer the bottom plate layout to this line. Then we pull a string through the stud layout of both the bottom plate and this line. We mark the points where the string crosses the line of the top plate.

Once the layout is accurately transferred to the top plate, it's a simple matter to measure stud lengths. We refer to the measurement as being to the "long side" or the "short side" to avoid confusion, and if we're building more than one wall from the pattern, we write the measurements on the deck below each stud.

In the air—When there is no deck on which to lay out the wall, we take a different approach. If the wall is short enough that the top plate can be cut from a single piece of lumber, we simply cut the plates and the end studs, nail these together and erect this frame. Once the frame is up and braced plumb, we lay out the studs on the bottom plate. If the wind isn't blowing, the layout can be transferred to the top plate with a plumb bob (photo right).

If the wind *is* blowing, we nail a 2x4 horizontally across the outside of the frame, level with the top of the shortest stud. We then transfer the stud layout onto this. Next we stretch a string from the bottom plate layout, through the layout on the 2x4, to the top plate and mark where the string crosses the top plate. If we don't trust the straightness of the top plate (and we never do), we pull a string along its top and use a temporary stud to correct the bow.

If the wall is long enough to require a two- or three-piece top plate, the wall can be equally divided. The lengths of the studs that will stand beneath the breaks in the top plate can be determined easily with a little math.

Let's return to the hypothetical wall: the length is 12 ft., the shortest stud is 8 ft., the longest stud is 14 ft., and the top plate is 13 ft. 5 in. If we were to break the top plate into two equal pieces 6 ft. 8½ in. long, the length of the stud that would stand under this break in the plate would be equal to half the difference between the length of the shortest and the longest stud, plus the length of the shortest stud, or 11 ft. Again, it helps to use the triangle/rectangle analogy. Simply put, if you cut the triangle in half, the legs will also be cut in half (bottom drawing, facing page). This works with any division.

When this method is used, the wall usually ends up with an extra stud in the center because the layout almost never coincides with the exact center of the wall. Sometimes the center stud interferes with the installation of another stud, but still doesn't fall on the layout. In this case, we simply add another stud onto the side of the center stud closest to the layout. Keep in mind that if you want the break to fall in the center of the stud, the measurement you arrive at mathematically will be to the center of the angled cut at the top of the stud. This is the only time we deal with a measurement that is neither the "long side" nor the "short side."

Opposite sides of a peaked wall should be identical, and when building "in the air," mea-

When there isn't room on the deck, Sheehan's crew builds rake walls "in the air," standing up the basic frame, then filling in the studs. After laying out the bottom plate, they use a plumb bob to transfer the layout to the top plate, provided the wind isn't blowing too hard.

surements can be transferred from the top plate on one side to the top plate of the other. If things start coming out "a little off," find out why. Geometry is an exact science, and if the studs that fit on one side are suddenly ¼ in. too long on the other, resist the temptation to just squeeze them in, or push them over and figure it's close enough. Chances are good that ¼ in. isn't nearly close enough.

It's very important to maintain close tolerances when balloon-framing. Particularly on steep pitches, an error in stud length of ⅛ in. can cause a considerable bow in the top plate. This is also true with regard to placing studs on center, and to a lesser extent, it holds true with top plate length. An error of ½ in. can really throw things out of whack. Double-check your math to be sure the figures are right. Once the wall is up, we usually tack the bottom plate

in a few places, brace it and immediately plumb it. When the ends are plumb, we run a quick check on each stud with a 5-ft. level.

Finally, when the wall is permanently nailed and braced, we snap a line the length of the wall at 8 ft. from the floor to serve as an installation reference for the fire blocking. In the event of fire, this will prevent a wall cavity from behaving like a chimney and increasing both the rate of spread and intensity of the conflagration. We position the blocks in an alternate pattern: one above the line, the next below, so we can nail through the studs and into the blocks. There is a danger of bowing the studs with over-dimension blocks, so here again, we maintain a high standard of accuracy. □

Sean Sheehan is a builder in Basin, Montana. All photos by the author.

Framing Walls

Speed and efficiency are the results of a careful, well-thought-out layout

by Scott McBride

Plated and detailed. Information written on the wall plates and the girders defines how the walls are put together. The author also lays out the second-floor joists or roof rafters before assembling the walls. Different-colored ink is used to denote a change in stud length. (Penciled guidelines were drawn on the lumber only for the sake of neatness in this article. It is not common practice for the author.)

I've heard a lot in recent years about the speed and the efficiency of California framers, but I find it hard to imagine anyone faster than the Italian-American carpenters who taught me to frame walls in the suburbs north of New York City. These men worked with an extraordinary economy of motion.

I want to discuss wall framing in general and, more specifically, to point out some of the methods and tricks I learned while working with New York carpenters. Even though some of these framing methods differ from those practiced elsewhere in the country, they have worked well for me, and I think they can work for anyone who wants to be more efficient on the job.

Carpentry has a vocabulary all its own. Stud, jack and header all have meanings outside the carpentry world, but to a framing crew these terms have specific definitions as components of a wall. If you are confused by a sentence that reads, "Toe-nail the king stud to the bottom plate," then you should familiarize yourself with the drawing on the facing page.

Snapping chalklines—The first step in any wall-framing method is snapping chalklines on the plywood deck to indicate the locations of the various walls. Wall locations will be shown on the plans. First, I snap lines for all the exterior perimeter walls. If I'm building 2x6 exterior walls —which I usually am these days—I use a 2x6 block to gauge a mark 5½ in. from the edge of the deck at each end of each my

2x6 template. A 2x6 block is used to mark an exterior wall's width on the plywood deck. The author sights down to the floor below to make up for possible irregularities in the alignment of the floor joists.

the block, I sight down to the corner of the foundation or of the story below, aligning the outside edge of the block with this vertical line of sight (photo below). You can't depend on the rim joist (called the box beam in New York) for registering the block because the rim joist is often warped out of plumb. Staying in line with the true corner is desirable, even if it means a bump in the sheathing at floor level. Otherwise the building tends to grow as it goes up, causing inconsistencies in the span that can complicate the roof framing.

After making a mark on all the corners of the deck at 5½ in., I connect the pencil marks with chalklines. I anchor the end of my chalkline with an awl tapped into the deck. When all the exterior walls are snapped out, I move on to the interior partitions, taking measurements from the plans and transcribing the lines onto the deck. I snap only one line for each wall and scrawl big Xs on the deck with my lumber crayon. The X indicates the side of the line where the wall goes. If there are 2x6 interior partitions as well as ones made of 2x4s, I indicate with my crayon which partitions are which.

Plating the walls—Plating is the process of cutting to length the bottom and top plates of the walls and temporarily stacking them on the deck (photo above). They can then be marked up to indicate where the various studs and headers will get nailed. In essence, I temporarily put all of the walls in place without the studs in them. My

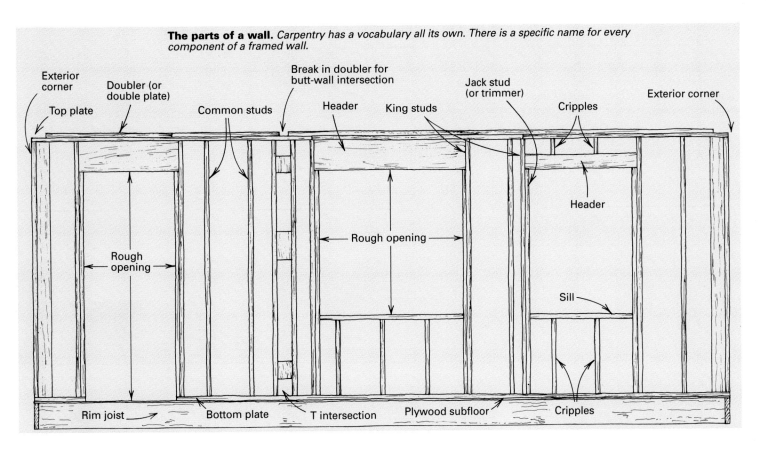

The parts of a wall. *Carpentry has a vocabulary all its own. There is a specific name for every component of a framed wall.*

method of framing differs a little from some others in that I cut the doublers now and stack them on top of the other two plates for a three-layer package. Later I'll explain why I do this.

Before cutting any lumber, I think a little about the order in which I want to raise the walls because this sequence determines how the corners of the walls should overlap. Where walls intersect, one wall runs through the intersection. This is called the bywall. The other wall ends at the intersection. This is called the butt wall.

By-walls have bottom and top plates of the same length that run through the wall intersections. The doubler of a bywall is shorter than the top and bottom plates by the width of the intersecting wall's plates (drawing above). This allows the doubler from the intersecting butt wall to lap the top plate of the bywall. Nailing through the doubler at the lapped corner into the top plate of the intersecting wall holds the walls together.

It is a good idea to cull through your lumber and save your straightest pieces for the longest top plates and doublers. I use the next-best stuff for the bottom plates, which are easier to straighten by nailing to the subfloor. The crooked stuff I cut up for short walls.

There are a couple of things to keep in mind when you are cutting to length the plates and the doublers. Butt joints in the bottom plate can occur almost anywhere. Splices in the top plate—the middle layer—should be offset as much as possible from adjoining walls and beam pockets. Here's why: The integrity of the top-plate assembly—the top plate and the doubler nailed together—depends on having well-staggered joints. An interruption of the doubler is inevitable at wall intersections and beam pockets, so keeping the joints in the middle layer *away* from these

points will maintain good overlap and avoid a weak spot. Splices in the doubler should be kept away from splices in the top plate by at least 4 ft.

If two walls cross each other, you'll have to let one of them run through the intersection and separate the other into two butt walls. The butt-wall doublers can split the overlap, with a joint in the middle of the bywall. Another option is to let one of the butt walls overlap in a full conventional tee. The other butt wall gets no overlap, but instead it is tied to the intersection with a sheet-metal plate on top of the doublers after raising the wall.

To commence plating, the bottom plates are toenailed to the deck on the chalkline, using 8d common nails about every 8 ft.—just enough to hold them in position. The top plates—the middle layers—are then temporarily toenailed to the bottom plates. Finally, I lay down the doublers—the third layer—over the top plates, but instead of tacking them temporarily, I nail them home with 10d common nails staggered 24-in. o. c. I use 10ds here instead of longer nails because the 10ds won't penetrate the bottom plates.

Remember, at the corners where the walls meet, the orientation of the butt joints is reversed, creating the overlap in the doublers that will ultimately lock the walls together. I don't nail the overlap now because I'll have to separate the walls later.

Detailing the plates and the doublers—
When all the plates are laid down and held together, and the doublers are nailed in place, I'm ready to mark them up, or detail them, with the information my crew and I will need to frame the walls. The first information recorded on the doublers is the width of the door and window rough openings in the exterior walls. The rough-

opening marks I make on the top of the doubler are discreet, only about 1½ in. long. I'm saving most of that surface for a later step in the layout.

If windows or doors are shown on the plans dimensioned to centerlines, I measure from the outside corners and mark the centerlines on the outer edge of the doubler. Then I divide the rough-opening dimension of the door or window in half. For example, if the width of the rough opening for a pair of French doors is 6 ft. 4 in., I'll align 3 ft. 2 in. on my tape with the centerline, then mark lines at 0 and 6 ft. 4 in. To check my arithmetic, I turn my tape around end for end. The center should still be at 3 ft. 2 in. I make a V to indicate the rough-opening side of each line (top photo, facing page).

Rough openings for interior doors are marked the same on the interior-wall doublers as for exterior doors, but the plans will usually call out the size of the finished door rather than a rough opening. To find the rough opening of a door, I add 2 in. to the width. This allows for a ¾-in. thick jamb and a ¼-in. shim space on both sides.

After locating a rough opening on the doubler, find the length of its header. Openings of less than 6 ft. will require one jack stud, or trimmer, on each side of the opening to support the header. Each jack stud is 1½ in. thick, so the header needs to be 3 in. longer than the width of the opening. (Because of variations in the actual thickness of 2x stock—studs can vary in thickness from 1⅜ in. to 1⅝ in.—it is a good idea to measure the lumber you're working with, then do your addition.) Headers over 6 ft. long require double jack studs on each side. That means the length of the header must be 4 times 1½ in., or 6 in. longer than the width of the rough opening.

I use a 2x block as a template to mark the jack locations on the outside of the rough-opening marks. I square the outermost mark down across the stacked edges of the three layers of 2x. This line indicates the end of the header and the inboard face of the king stud. The king stud is the full-length stud to which the jack stud is nailed. On the edges of the top and bottom plates, I show the king stud with an X and the jack stud with an O. For double jacks I use OO. After repeating the process on the other side of the opening, I measure between the outermost marks on the top of the doubler to verify the header length I arrived at earlier by arithmetic. And finally, I write the length of the header on the doubler.

Window headers are marked out the same way as for doors. As far as windowsills and bottom cripples are concerned, I usually come back to them after the walls are up. They aren't needed for structural reasons, and I'm usually in a big hurry to finish the framing and get the roof on. But if I'm going to sheath the exterior walls before they are tipped into place, I frame below the windows as I go. In that case I'll write the height of the window rough opening on the doubler as well as the width.

I think presheathing pays if you can have the plywood joint even with the bottom of the wall. But some builders, architects and inspectors require that the plywood joints be offset from the floor elevation to tie the stories together. You can still presheath in that case by letting the sheets hang over the bottom plate, but I think it becomes more trouble than it's worth. It's usually more economical to let the least skilled members of the crew hang plywood after the walls have been raised.

As soon as I've finished determining all the header sizes, I make a cutlist and give it to a person on the crew who can then get busy making headers while I finish the layout.

Detailing the doubler—After marking the rough openings on the edges of the top plates and the doubler, the focus shifts back to the top face of the doubler. It's time to lay out the structure that will eventually sit on top of the wall you're about to build. There are very logical reasons to do all this layout now. First, by doing it now, you won't have to spend a lot of time working off a stepladder or walking the plate after the walls are up, and the doublers are 8 ft. in the air. The work will already be done. Second, it is easier to align the studs in the wall you're currently building with the loads coming down from above. This is called stacking, and I'll discuss it further later in the article.

The structure that sits on top of the doubler may be either a floor or a roof. In either case I start by locating the principle members—girders in the case of floors, ridges in the case of a roof. I mark their bearing positions on the doublers. Then I measure their actual lengths, cut the members and set them on the doublers. I can now lay out the spacing for the joists or the rafters on the principle beams at the same time I put the corresponding layout on the walls.

To ensure that the principle member is positioned correctly when it's eventually raised, I make a directional notation on the beam, such

Layout trick. A flexible wind-up tape is used for layout. Hold a pencil against the tape and swing it across the plate to make marks. The farther the tape is away from the end of the plate, the straighter the lines will be.

Headers and king studs. Tipping the nailed-together doubler and top plate upside down on the deck makes it easy to toenail the header and attach the king stud.

Scribing the jacks. By holding a stud alongside the king stud and against the underside of the header, the jacks can be scribed without measuring. Visible on the bottom plate, half hidden by the carpenter's hand, is a galvanized plate nailed over joints in the plate.

as north end, or driveway side. With the principle members in position, the first joists or rafters I locate are specials, such as stairway trimmers, dormer trimmers and joists under partitions.

When all the specials have been marked, I lay out the commons. These are individual, full-length framing members (either joists or rafters in this case). The standard spacing is 16 in. o. c., but 12 in. and 24 in. are not uncommon. You want to minimize the cutting of your plywood, so the spacing of commons should result in 8 ft. landing in the middle of a framing member. Assuming 16-in. centers, I hook my tape on the end of the doubler and tick off 15¼, 31¼, 47¼, etc. As an alternative, you can tick off the first mark at 15¼ and then measure the remaining marks from there on exact 16-in. centers.

You can square the tick marks across the doubler with a square or use this trick: Clamp your pencil point against your tape at the given spacing by pinching them together between your thumb and fingers. Swing an arc across the plate (top photo, left), move your pencil to the next point, repeat and so on. The error caused by the curvature of the line is negligible, and it diminishes as you move outward. This trick works better with a flexible wind-up tape than with the stiffer, spring-return tape.

By now you're probably wondering if I'm ever going to start nailing the walls together. Be patient, we're almost ready.

Stacking studs—The spacing for the common studs in the walls is derived from the joist or rafter spacing in the structure that sits above. If joist or rafter spacing is the same as the stud spacing, both 16 in. o. c. for example, I simply extend the layout mark on the top of the doubler down across the edges of the plates. Aligning the framing members like this is called stacking. But if you have, say, joists on 12-in. centers and studs on 16-in. centers, every fourth joist should stack over every third stud. Stacking wherever possible helps to prevent deflection of the top plate, although it could be argued that the presence of sheathing and band joists makes such deflection unlikely. Stacking *does* facilitate the running of plumbing pipes as well as heating pipes or HVAC ductwork.

And on exterior walls, stacking studs from one story to the next makes installation of plywood sheathing more efficient. To support a special joist, such as a double 2x trimmer, I square the layout mark on the doubler down across the edges of the plates and indicate a special double stud.

Over door-header locations, I extend my pencil marks for the cripple-stud spacing across the edge of the doubler and the top plate, but I stop short of the bottom plate. Instead of marking an X here, I mark a C on the correct side of the line. This will show the carpenters where to put the cripple studs above the door header. Of course this step is unnecessary if the header sits tight against the top plate. At window headers I extend the mark across the bottom plate as well, writing C for cripple.

The studs in nonbearing partitions running perpendicular to the joists should also stack, although it's not as crucial as for bearing walls. (A nonbearing partition is a wall that doesn't support a load.) For partitions running parallel to the joists, the placement of the common studs is discretionary. They can be laid out from either end of the wall.

Building corners—With the stud layout completed, there are only a few more details that need to be mopped up before nailing the walls together. At the end of the exterior bywalls I write CORNER, which means a U-shaped corner unit made up of two 2x6s and one 2x4. This corner design permits easy access for fiberglass insulation (top drawing, right).

Another step in the final layout is to mark out the studs that go at the end of each butt wall where they intersect a bywall. These inside corners provide nailing for drywall and baseboard and, for that matter, any other type of finish trim that might end in a corner.

Some carpenters preassemble channels to back up these T-shaped intersections. I find it easier to space a pair of studs in the bywall, separated by the flat width of a block. I add the blocks after the walls are raised: one in the middle for 8-ft. walls, two or more for taller walls. By using 2x6 blocks behind 2x4 partitions, and 2x8 blocks behind 2x6 partitions, I get a 1-in. space on both sides between the partition and the bywall studs. This provides access for insulation. It also makes nailing drywall and baseboard easier because you don't have to angle the nail as much to catch the corner stud (middle drawing, right).

Corner posts for interior 2x4 partitions are made up of intermittent blocks that are sandwiched between full-length studs. This type of corner can also be used at the end of peninsular walls (bottom drawing, right). Write B to indicate the blocking.

The final layout step is to number each wall for identification and to indicate the raising sequence. As a convention I write the number on the left end of the doubler as I look down on it. I then write the same number in front of it on the deck in heavy crayon. These steps help prevent the wall from being installed backwards, which is easy to do. I put a slash under the 6 and the 9 to tell them apart. If there are different stud lengths within a story, I write the appropriate length next to the raising-sequence number that has been assigned to each wall. It's not a bad idea to use a different-colored crayon or marker to indicate different nonstandard stud lengths. For example, if I'm writing everything else out in black crayon, I use a red crayon to make the exceptions easy to spot.

Nailing the walls together—When the layout detailing is complete, the temporary toenails are removed from the bottom plate only. Each wall is now represented by a separate package containing the bottom plate, the top plate and the doubler. I stack these packages in an out-of-the-way place on the deck, along with the headers, the corner units and the principle beams for the structure above. Studs should be leaned against the edge of the deck where they can be reached, but not stacked *on* the deck. The fastest

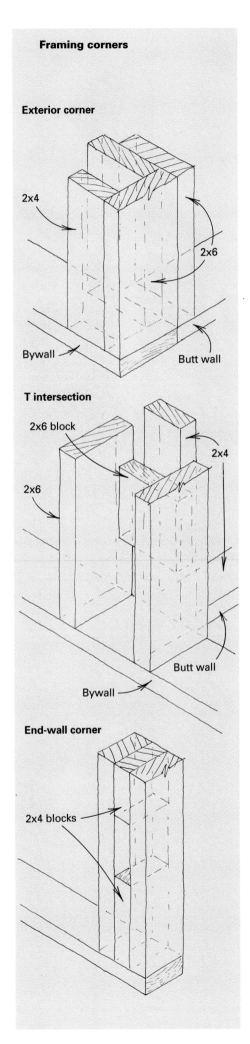

Framing corners

Exterior corner

2x4

2x6

Bywall

Butt wall

T intersection

2x6 block

2x4

2x6

Butt wall

Bywall

End-wall corner

2x4 blocks

and most accurate way to mass-produce non-standard studs is with an improvised double-end cutoff arrangement. Nail two chopsaws or slidesaws to a bench at just the right distance apart. Two operators working together lift a stud onto the beds and cut off both ends. This method squares up both ends of the studs and cuts them to length.

When I'm nailing walls together, I usually start with one of the longer exterior walls. I lay the plate package on the deck, parallel to its designated location and pulled back from the edge of the deck by a little more than the length of the studs. You don't want to crowd yourself. I pull out the nails that hold the bottom plate to the top-plate assembly and spread the plates, moving the bottom plate close to the edge of the deck. I'm real careful not to turn the plate end for end.

I find the wall's headers and carry them over to their locations. If the header sits tightly against the top plate, I flip the top-plate assembly upside down and toenail the header down into the underside of the top plate. Then I stand a king stud upside down on the plate and through-nail it to the end of the header with 16d sinkers or 10d commons. I throw a few toenails through the king stud down into the plate as well (middle photo, facing page). Now I roll the assembly down flat on the deck.

Some people precut all their jacks, but I like to make them as I nail the walls together. It's simple and fast: I take a common stud and lay it against the king stud, one end butted tightly against the header. Then I strike a line across the bottom end of the king stud onto the jack (bottom photo, facing page) and cut carefully, just removing the pencil line. I nail the jack to the king stud in a staggered pattern, 16 in. o. c. This method of cutting jacks in situ compensates for the variations in header width. The short cutoffs will be used up quickly for blocking.

If the header is offset from the top plate by cripple studs, the wall-framing procedure is substantially the same, except that it's all done flat on the deck, toenailing cripples to the top plate and then to the header (photo p. 48).

Toenailing the studs—Because the top plate and the doubler are nailed together beforehand, the tops of the studs must be toenailed in place rather than through-nailed. Toenailing requires more skill than through-nailing; it might take a beginner a little longer to learn, but it's not as if it's something he won't have to learn eventually. And toenailing is stronger than through-nailing because it penetrates across the grain. As the walls are jockeyed around on the deck and moved into position to be raised, the bottom plates, which are through-nailed, loosen easily while the toenailed tops hold firm.

If you'd rather not toenail, or if you're using air nailers, which make toenailing difficult, you can tack the doubler to the top plate temporarily for layout purposes. Then pull the doubler off to through-nail the tops of the studs. And finally, nail the doubler back in place using an index mark to ascertain its correct position. Toenailing is the method I was taught years ago, and it's

Nailing the walls together. After all the layout and cutting is finished, the wall components are nailed together. The author nails the doubler to the top plate before the wall is assembled. This then requires that the tops of the studs be toenailed to the plate, instead of through-nailed as in other methods. The pull-out strength of a toenail is greater than that of a through-nail into end grain.

what I'm most comfortable with, so that's the method I'll describe.

When the headers and their jacks and king studs have been nailed in place, I stock the wall with common studs. One end of each stud rests on the top-plate assembly so that it won't bounce around when I start my toenails. I start at one end of the wall, lift up a stud, quickly eyeball it and lay it back down with the crown pointing to the left (because I'm right handed). I work my way down the length of the wall until I reach the end (photo right). For 2x6 walls I use 10d commons or 16d sinkers for toenailing. For 2x4 walls I use 8d commons or 10d sinkers. Starting at one end, I start my toenails in the upturned face of each stud—three nails for 2x6, two nails for 2x4. Ideally, the point of the nail should just peek through the bottom of the stud. I work my way down the row.

Bracing the top plate with my feet, I grab the first stud in my left hand. As I shove it away, I turn it 90° to the right so that it lies on edge, then I pull it back up firmly against the plate. Because the crown now faces up, the stud won't rock on the deck. One blow sets the nail, and two or three more drive it home (photo facing page). The stud will drift as it's toenailed, depending on the accuracy of the cut, the accuracy of the hammer blow and the hardness of the wood. Even if you are just a beginning carpenter, you'll quickly learn how far off the mark to start as a way of compensating for the force of your hammer blows.

Some carpenters drive a toenail in the 1½-in. edge of the stud to hold it in place, but I've always thought this was a superfluous practice that can be dispensed with. When I reach the end of the wall, I double back, firing nails into the other side of the stud—two nails for a 2x6 stud, one nail for 2x4, staggered with those on the other side.

When the studs, the jacks and the cripples are toenailed to the top plate, it's time to nail the bottom plate. The bottom plate gets through-nailed to the stud with 16d sinkers—three to a 2x6, two to a 2x4 (photo below right).

Raising the walls—There are two schools of thought regarding the sequence in which the various walls should be raised. If space on the deck is tight, the long walls must be framed and raised before the other walls have been inflated with studs. Otherwise there won't be enough room. But if I have some room to spare, I'll start with the littlest walls.

I frame the walls and start piling them up. When the pile is finished, the little walls will be on the bottom, and the medium-length walls will be on top. Finally I frame and raise the long, exterior walls. Instead of bracing the long walls with a lot of diagonal 2x4s attached to scab blocks, I can pull an adjoining medium-length partition off the pile and drop it into place. This immediately buttresses the long wall. As I work from the perimeter of the house toward the inside, the pile diminishes, and the walls pop up quicker than dandelions.

It may go without saying, but when raising walls, especially long, heavy 2x6 walls, it is important to lift using the power in your legs rather

Crowning and toenailing. Sighting down each stud determines its crown. All the studs are then laid on the top plates with the crowns facing the same direction. The toenails are then started in all the studs and nailed all at once, production-line style.

than the smaller muscles in your back. After a wall has been lifted into a vertical position, I ask one man on the crew to line up the bottom plate to the chalkline on the deck while the others hold the wall steady.

The wall can usually be moved to the line by banging on it with a sledgehammer. I nail the bottom plate into the floor framing, one nail per bay. A bay is the space between studs.

All that remains is to rack the walls plumb and brace them (see pp. 58-61). The only braces sticking out into the room should be those necessary to straighten any long, uninterrupted walls that are crooked. The rest of the walls will have been braced by each other. With the next phase of the job already laid out, you're ready to rock and roll. □

Scott McBride is a contributing editor of Fine Homebuilding. *Photos by Charles Miller.*

Through-nailing. As in other methods of framing, the author prescribes through-nailing the bottom of the studs, the jacks and the cripples. As he works his way down the plate, he aligns each component to its mark and nails it.

A crowbar provides leverage and control. One person moves the wall; the other tells when the wall is straight up and down, or plumb. Here, a crowbar forces a racking brace forward, moving the top of the adjacent wall. A block nailed to the bottom plate provides even more leverage. When the wall's plumb, the brace is nailed to the bottom plate.

Plumbing, Lining and Bracing Framed Walls

A framing contractor explains his efficient sequence for keeping walls straight and true

by Scot Simpson

I have a painting on the wall of my office at home. The painting hangs in a room built in the 1920's or 30's, and the room has 2x4 floor joists and rafters. No matter how hard I try, I can't get the picture to hang straight. My office was probably plumb and straight when it was new, but now it's neither straight nor square nor level. A picture that doesn't hang squarely on a wall might not bug you, but even worse things can go wrong if you build an out-of-square structure.

To avoid these problems, once you've built, stood and nailed walls together, they must be plumbed and lined. Plumb and line makes walls straight and true. Plumbing is setting a level against the end of a wall to make sure that it stands up straight. Lining is using a tight line attached to the top of a wall to gauge a wall's straightness along its length. You plumb and line walls before installing joists, rafters and sheathing because it's nearly impossible to move walls after these items are nailed in place.

Doing an inaccurate job will slow down every subsequent framing phase. If exterior walls aren't straight, you'll have to measure every rafter or floor joist before you cut it; otherwise, it won't fit on the crooked wall. And if you have crooked walls, you'll have bowed siding; you'll have to scribe soffits. You'll have a mess.

If you've framed the floor and the walls carefully, it shouldn't take more than a few hours to plumb and line the walls. You can speed the process by approaching the tasks methodically and by eliminating unnecessary steps.

Start with careful layouts and straight lumber—When framing walls, you should use the straightest lumber possible for the top and double-top plates, the corner studs and the end-wall studs. And make sure you cut the bottom and the top plates exactly the same length.

Once walls are stood, be sure all intersecting walls are nailed together tightly and that all double-top plate laps are tight. Intersecting walls must line up in their channels, and corner studs must line up.

Next, bring in your brace lumber—an assortment of 2x4s ranging from 8 ft. to maybe 14 ft., depending on wall height—and distribute the braces throughout the house. These braces help you move walls into position and hold them there until the joists, rafters and sheathing go on.

To plumb and line, you first get exterior corners standing straight up, or plumb. Next, you straighten the top plates so that they're in line with the walls' corners. Then, you plumb and line any interior walls that intersect exterior walls and finish by plumbing and lining remaining interior walls.

Plumbing exterior corners—Pick any exterior corner of the building as your starting point to plumb the walls, but work in one direction after that because when you move one wall, you also move any walls attached to it. Take care not to move walls that you've already plumbed.

You'll need two people—one who will move the wall, the other who will check it for plumb (photo, facing page). The person who's checking the wall puts a level against the end of the wall to see which way it's leaning. I use an 8-ft. level or a 4-ft. level attached to a straightedge that reaches from the top plate to the bottom plate.

When let-in braces are required, you can use a push stick to square up the walls (photo, right). A push stick is usually just a 2x4 that you place against the face of a stud toward the top of the wall. The bottom of the push stick rests on the floor. I bought a bracket (Rack-R, P. O. Box 974, Novato, Calif. 94948; 415-897-7044) that fastens to the bottom of a push stick and keeps it from kicking out. You lean or even step on the push stick to move the wall into plumb. When the person on the level says right on, the let-in bracing is nailed off, pinning the wall plumb.

Unfortunately, a push stick doesn't produce enough force to move rigid walls. Plus, you still need a brace to hold the wall in place.

Because the walls I frame have sheathing, there's no need for let-in wall braces. But I still have to push walls plumb and hold them there until the sheathing goes on, so I use a racking brace. A racking brace is a 10-ft. or 12-ft. 2x4 placed diagonally on the face of a wall. The high end of the brace points in the direction the wall needs to be pushed.

You nail the brace to the top plate. If you're nailing by hand, you probably will want to start two nails in the end before you lift the racking brace into position. Make sure the brace doesn't stick above the wall in the way of joists or rafters.

The bottom end rests on the floor, and you use a crowbar to shove the brace forward, which pushes the wall. A brace installed at a 45° angle gives the best leverage; a steeper angle tends to push the top plate up.

A metal fitting lets you work alone. A push stick—a relatively knot-free 10-ft. 2x4—is used to nudge a wall plumb, but it doesn't supply as much leverage as a racking brace and a crowbar. The metal fitting at the bottom of the push stick keeps the brace from slipping.

When the person on the level says the wall is plumb, the brace person drives one nail through the brace into the bottom plate. Then, he releases the crowbar from the brace while the other person checks both ends of the wall for plumb. If both ends are good, then a second nail is put into the bottom plate and one nail is set into a stud in the middle of the brace.

Move to the next corner down the line. Move in one direction until all exterior walls are plumb.

Sometimes, the wall doesn't want to rack with a crowbar, but hitting the double-top plate at the

end of the wall with a sledgehammer tends to loosen it up. Also, you can nail a block behind the crowbar to increase leverage when forcing the racking brace.

Lining exterior walls—Now, you've got all exterior corners plumb. Next, straighten the walls. I use two methods to straighten walls. First, I rack any interior walls butting into the exterior walls. When these interior walls are plumb and square, the walls they butt into should be straight. I say "should" because sometimes it just doesn't work

Blocks make sure the line stays straight. A traditional method of lining is to nail blocks at each corner and to string line across the blocks. Blocks hold line away from the plate. Another block gauges distance between string and plate.

Tying off a line. To secure the line on the nail, pull the string tight, twist your finger around the string thrice and slip the loop over the nail. Then, just wrap the loose end around the nail.

Using a line without blocks. You can tell if a wall is straight by stringing a tight line from one end of a top plate to another. A nail is bent flush with the outside plane of the wall, and another nail is tacked to the opposite corner.

that way, and you've got to find the problem and fix it. I'll get to how you fix it a little later.

In places where there are no interior walls to rack and push the exterior walls straight, I use line braces (photo, right) to push or pull the top plate into alignment. A line brace is usually a 2x4 face nailed to a stud just beneath the top plate and spiked to a cleat on the floor or to another wall's bottom plate.

Line braces hold the top of a wall in place like a prop, and they keep the wall straight and stable so that it's safe to walk on as you're nailing off joists. Even if the wall is straight, these braces should be placed every 10 ft. along the wall and at breaks in the top and double-top plates, which are typically weaker points. Make sure the line brace doesn't interfere with wall sheathing. Racking braces and line braces usually stay in place until the roof has been framed and the walls and roof have been sheathed.

Using a stringline—Some carpenters determine where to straighten a wall by sighting along the top plate the way you'd check a board to see which way it crowns. I don't like to sight the plates because it's not always accurate. Often, the edges of the plate lumber are waned: curved instead of square due to bark or defects. And it's easy to be confused by the many lines in a house's frame, which can cause optical illusions.

For a better look at how straight the top plates are, I run a line from one corner of a wall to another, pulled tightly between nails at each end of the top plate. The line tells me how far in or out the wall is, and by matching the top of the wall with the line, I can straighten the wall.

Commonly, lining is done using blocks. You

Line braces hold walls straight. This gable end is straightened and held straight with 2x4 line braces. Note the line running along the blocking. The top of each brace is nailed to the wall just below the plate. The bottom of each brace is left square so that it can be pushed forward with a crowbar, just like a racking brace. When the wall is straight, the line braces are nailed to cleats. The cleats are spiked to the floor joists, not just to the plywood deck.

face nail a piece of 2x lumber at the top corners of a wall, and then string a line on the outside of the blocks (photo above, left). Then, you slide a 2x4 block along the top plate, checking to see where the block pushes against or pulls away from the line. These are the spots where the plate waves in or out, so you'll need to set a line brace or rack an intersecting wall to straighten the top plate. The purpose of the blocks is to hold the line away from the wall so that the line remains perfectly straight regardless of imperfections in the plates. Without blocks, a crooked plate touching the string somewhere along its length would make the string crooked.

For years I used the blocking method. But I once tried a different method, and I haven't gone back. The line method is the better system. It eliminates the extra steps of putting on and taking off the blocks. It also eliminates the problem of getting an accurate reading if the double-top plates aren't exactly flush. With no blocks, you see the line in relation to the whole wall, not just a protruding edge on the top plate.

To set the line, start a 16d nail in the top of the double plate near the end and edge of the wall (photo above, right). Then, bend the nail so that the string will be in line with the edge of the top plate and wall below. Cinch a string to the nail about ½ in. above the plate, extend the string to the other end of the wall and set another nail. Attach the string, pull tight and secure. The easiest way to secure the string is by placing it around your finger and twisting it three times (photo above, center). Then, put the string over the nail. Use one hand to pull the string coming onto the nail while the other hand pulls the end of the string coming off the nail. Once the string is tight, wind the loose end around the nail and tie it off so that the coils trap the loose end.

The key to the line method is that the string is in line with the edge of the wall but is ½ in. above the plate. Hence, no blocks are necessary because imperfections in the plate won't touch the string. And with the plate and string so close to each other, you visually can compare the top plate to the string, and you can adjust the plate accordingly.

To straighten exterior walls, attach a racking brace to each interior wall that runs into the exterior wall. Rack each interior wall until the lined wall is straight, then nail the brace. If an interior wall runs between two exterior walls, you have to line both exterior walls before racking the interior wall with a racking brace.

For sections of wall that do not have walls running into them, place line braces wherever necessary to hold the wall straight.

If the top of a wall leans into the house, I use a crowbar to shove the bottom of a line brace forward and push the wall out.

If the top of a wall leans away from the house, I first try to pull the wall in line by hand with a line brace. If the wall won't budge, I use a pulling brace (photo, right). This brace is pretty much like a line brace, except a pulling brace is installed on the flat instead of on edge, and it draws the top of a wall in instead of pushing it out.

The top end of a pulling brace is nailed under the top plate. The bottom end is trapped under a pair of blocks nailed to the floor. I usually nail the blocks perpendicular to the brace, with the top block offset from the bottom one so that it holds the brace down. The object is to bow the pulling brace upward—which shortens it and draws the top of the wall in—and the blocks hold the brace down better than if nailed to the floor.

To bow the pulling brace, I nail the bottom of a 4-ft. 2x4 block to the floor, jam the top of the block under the brace and pull the block tighter and tighter under the pulling brace. If the wall's stubborn, I bash the block under the brace with my hammer. Eventually, the block should bend the brace enough to pull the wall into line; at that point I drive a few nails through the pulling brace into the top of the block, which holds the wall in place.

Walls with long headers are notoriously difficult to line, and sometimes you need more oomph than you can get from a pulling brace. That's when I break out a come-along.

Making adjustments—If the walls were built perfectly and the floors were all level, you shouldn't have to make any adjustments to your plumbed and lined walls. However, when you're dealing with lumber that varies in dimensions and often is warped and bowed, and when the plate's end cuts may not be exactly square or the sections of plate are not nailed tightly together or kept tight, it's not unusual that some trimming or stretching becomes necessary to get everything straight and plumb.

The first thing to do when an adjustment is needed is to find the mistake. I check the typical trouble spots. First, I check for a tight fit between intersecting top plates. A gap here usually

Pulling braces need to be anchored to the floor. When a top plate has to be pulled in, you need a pulling brace, a 2x4 nailed on the flat under the top plate. A short 2x4 then is used to bow the brace upward, hence pulling in the wall's top plate.

can be corrected by toenailing a 16d nail up through the top plate of the bywall into the top plate of the intersecting wall.

Then, I check that the top and the bottom plates are the same length. Sometimes there are spaces where plates lap; sometimes the plates are different lengths. If plates aren't the same size, you have to make the top plate longer or shorter.

The wall can be made longer with a few easy steps. Pull the nails where the plates lap. Then, plumb the corner, creating a gap in the top plates of the connecting walls. Nail the end stud of the one wall tight to the connecting stud of the other wall and renail the double plate. You'll have a

gap in the top plates at the corner, but the walls will be plumb.

The wall can be shortened by loosening the double plate from the top plate, cutting the top plate shorter with a reciprocating saw and nailing the top plates of the two walls tight. If the double-top plate overhangs the exterior, trim it. □

Scot Simpson, a framing contractor in Seattle, Washington, is the author of Framing and Rough Carpentry: Basics for Builders *(R. S. Means Co. Inc.). He is working with several trade associations on an international framing-training program. Photos by the author.*

Wall Bracing

Wood, metal and plywood can all be used to keep walls from racking

by Larry Haun

I was born and raised in rural Nebraska, where the one constant is the wind. Many of the farm buildings I remember from my childhood leaned to one side. I was fascinated by these buildings. I used to think that was the way people built them. It took a while to figure out that those buildings leaned because the wind was always pushing on their poorly braced walls.

The first house I helped build was in 1947, back in the handsaw days. For wall bracing we mitered the ends of 2x4 blocks, then we nailed the blocks between studs along a diagonal line—good braces to prevent racking but time-consuming to build.

I now live and work on the West Coast, where houses must be built to withstand not only the wind but also earthquakes. I've braced many a wall, and here I'll talk about the best ways I've found to keep walls plumb and true through gusts and tremors.

Let-in bracing—Once platform framing became popular, carpenters learned to brace walls by mortising, or letting in, a diagonal 1x4 or 1x6 into the plates and the studs of a wall (photo below). Many building codes call for this type of bracing in almost every wall that has room for it. Braces are required at each end of a long wall, with an additional brace for every 25 ft. of wall space. The codes don't say how long the braces should be; they just say you must have them. Each brace must oppose the other—one slanting one way, the next slanting the opposite way—to guarantee that one member will always be in compression (the force that pushes together or crushes). Let-in wood bracing works best in compression.

I like to use let-in bracing even if the wall will later be sheathed with plywood (another form of bracing). It takes me only two or three minutes to install a brace, and let-in bracing holds the building plumb until the sheathing goes on.

Other carpenters sheathe their walls before raising them or hold walls plumb with temporary 2x4 braces after they're up. But sheathing the walls first means squaring them up perfectly on the deck—a tedious job—and when temporary braces get in the way, workers tend to remove them: there go the plumbed and squared walls.

When shear forces go beyond the ordinary, a structural engineer can calculate how to brace a wall to keep it from racking and uplifting. Often, it's a matter of combining panels and seismic anchors (see sidebar on shear walls, p. 57). But where high winds or major earthquakes are uncommon, 1xs are adequate for bracing. For the inexperienced, installing 1x let-in bracing can be dangerous, so the procedure needs to be studied well and executed carefully to avoid injury.

Notching studs and plates—While the wall is still flat on the deck, and with both the bottom

Installing let-in bracing. **Walls are held plumb with let-in bracing: good-quality 1x6s mortised into the studs and the plates. The bottom of the brace is nailed to the bottom plate and the first stud only; the rest** **of the nailing will be done once the wall is raised and plumbed. Here the author nails the bottom of the brace while his brother notches studs for the next brace.**

and top plates nailed off, I get the wall relatively square by keeping the bottom plate parallel to its layout line and nailing the studs at right angles. It's not important that the wall be precisely square just yet. Then I pick out a good 12-ft. 1x6—#2 grade or better—and place it at about 45° on the wall studs from bottom to top plates. I don't run the brace end right into a corner; I keep it in a stud bay so that I won't hit a nail when I cut the notches.

When you install let-in bracing, keep in mind that the steeper the angle, the less effective the brace, so a 45° angle is better than a 60° angle, for example. On an 8-ft. wall, the 12-ft. 1x6 will cross over at least four studs (on 16-in. centers) and five stud bays.

I cut the end of the brace flush with the bottom plate. If I'm working on a concrete slab, I trim the bottom end of the 1x6 about ½ in. short to keep that succulent end grain away from termites. Next, I set my circular saw to a depth of about 1¾ in. I place one foot on the brace to keep it in position and hold the saw with both hands to avoid kickback. Now, with my saw riding on top of and parallel to the brace, I cut 1-in. deep kerfs in the studs and the plates (top photo, right). I cut on both sides of the brace, then cut the brace ½ in. short of the top of the double top plate. I cut the brace ½ in. short before raising the wall so that I won't have to trim the brace when the wall is plumbed. Cutting the brace short gives me a bit of play when plumbing a raised wall; otherwise, the brace may be long if the wall is not square.

Now the wood between the kerfs must be removed to accommodate the let-in brace. One way to remove the wood is to cut additional kerfs, closely spaced, to a depth of about 1 in. The wood can then be knocked away with a straight-claw hammer. But this method is slow and leaves an unclean notch.

The better way to make the notches requires care and experience to do it correctly and safely. It involves making a plunge cut into the sides of the studs between the two 1-in. deep kerfs. The object is to lower the blade into the stud or plate and cut out the piece between the kerfs so that the brace will fit. Just remember that the cut needs to be at least ¾ in. down from the edge of the stud. When you're making this cut, be sure to keep both hands on the saw. And when you're learning this technique, it's a good idea to brace your elbows against your knees so that you'll be able to stop the saw from kicking back.

Nailing the braces—Now I drop the 1x6 brace into the notches and nail three 8d nails through the brace into the bottom plate and two more 8ds into the first stud. That's it. I start two nails in the brace at each remaining stud and five more at the double top plate. These nails will be driven home when the walls are raised and plumbed (bottom photo, right). You can also start a nail in the top plate and bend it over the brace to keep it in position while raising the wall.

Rake walls and other tall walls can be braced by using a long 1x or two short ones. The short ones will lap in the center, which requires cutting a deeper notch in at least three studs. Even

Cutting kerfs. With the 1x6 lying diagonally over the relatively square wall, the brace becomes a template. One-in. deep kerfs are cut into the studs on both sides of the brace. The brace terminates at a stud bay, not at a stud, to avoid the possibility of hitting a nail with the saw.

with this deeper notch, I've never had an inspector question the compression strength of the studs. When plumbing the walls, I nail the lapped 1xs into the studs with 16d nails.

Metal-angle braces—Metal-angle framing braces (left drawing, p. 56) are fairly new in the construction industry. They're available at lumberyards, and the angle braces we use (CWB106 and CWB126 from Simpson Strong Tie, 1450 Doolittle Dr., San Leandro, Calif. 94577; 800-999-5099) are 18-ga. metal and come in two sizes: One is about 9½ ft. long and is installed at a 60° angle; the other is over 11 ft. long and is installed at a 45° angle. They offer holding power not only in tension, but because of the metal angle, they hold in compression also. Compared with 1x let-in braces, metal braces are faster, easier and safer to install and will hold the building plumb just as well.

To install a metal angle brace, I lay it across the framed wall diagonally just like a wooden 1x. I make a pencil mark on the plates and the studs along one side of the brace. Then I cut a kerf 1 in. deep along this pencil mark. I slip one flange of the brace into the kerf and nail the brace to the bottom plate with two or three 16d nails, then add one more 16d through the brace and into the first stud. At the double top plate, I start a nail alongside the brace and bend it over to hold the

Walls up; braces nailed. Let-in braces are nailed off after the walls are plumbed. The nails in the double top plate are driven first, followed by the nails in the rest of the studs.

brace as the wall is raised. Later, when the building is plumbed, the metal will be nailed permanently to the other studs and plates. The braces come predrilled: One flange has no holes; the other side looks like an incredibly uniform piece of Swiss cheese.

The 18-ga. braces I use are only about 1/16 in. thick, which causes no problem with siding. On taller walls, metal braces can also be lapped at the center. A slightly wider kerf has to be cut in the studs where the two braces overlap. I'll mortise the lapped section into the studs a bit to ensure there will be no lump in the siding.

Walls with large openings—The front wall of a garage, which is almost all opening, can be braced in several ways. One way is to plumb the walls and add a diagonal brace at ceiling height. I lay a 10-ft. or 12-ft. 2x4 diagonally across the double top plate from the wall containing the garage header to the sidewall (photo above). On a detached garage, I put the brace on both front corners. I mark the brace's location on the double top plates, cut out these sections and drop the brace into the notches. Once the wall is plumb and straight, I nail the brace in place with 16d nails and trim it flush with the walls.

A more traditional method of bracing garage walls is to use long or lapped 1x6s, nailed diagonally on the double top plates from corner to corner of the garage, forming a large X at ceiling

Bracing an open wall. Walls with large openings, such as the front wall of a garage, can be braced with a 2x4 let into the double top plate. Once the wall is plumbed, the 2x4 is laid diagonally from the open wall to the sidewall and fastened with 16d nails.

height. This bracing will hold the garage-door opening stable until rafter ties and ceiling joists are in place. The 1x braces can then be nailed up into the ties or joists for further stabilization.

A third method of bracing a garage wall is to nail a piece of plywood to the small section of wall on either side of the garage-door opening. Although these walls are generally quite narrow, perhaps 1 ft. or so, an 8-ft. high section of plywood will help brace the entire wall, especially if a let-in 2x4 brace has been installed up top.

Bracing walls with sheathing—Most walls are long enough to accommodate a whole sheet of

plywood or oriented strand board (OSB) on each end. Nailing one or two 4x8 panels on these walls should be adequate bracing in parts of the country where shear forces aren't much of a concern. Other panels nailed on short walls within a house, such as the backside of a closet or a bathroom, can help stabilize an entire building.

I usually start sheathing a plumb wall at a corner by driving two 16d nails between the bottom plate and the floor sheathing or concrete slab. I let these nails protrude an inch or so to support the panel, square the panel with the corner and tack it in place. A common nailing schedule for plywood or OSB panels is 8d nails at 4-6-12, meaning you drive a nail every 4 in. at the building's perimeter, 6 in. at joints between panels, and 12 in. in the field, or inner part, of the panel.

In colder parts of the country, builders may want to increase the R-value of the walls by sheathing the exterior walls with 1-in. foam insulation (right drawing, below). Wall bracing can be provided by nailing a sheet of 1/2 in. plywood at each corner and overlaying it with a sheet of 1/2-in. rigid foam insulation to bring it out flush with the 1-in. foam. □

Larry Haun is a carpenter who lives in Los Angeles, Calif. His book, The Very Efficient Carpenter, *and videos were released by The Taunton Press in 1992. Photos by Roger Turk except where noted.*

Metal-angle framing braces. *Metal-angle braces are easier and safer to install than wood braces because you only need to cut a kerf in studs and plates and drop the brace in. Metal braces are predrilled and are held with 16d nails.*

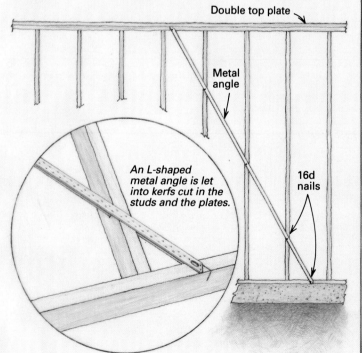

Double top plate

Metal angle

An L-shaped metal angle is let into kerfs cut in the studs and the plates.

16d nails

Panel bracing and rigid insulation. *When using 1-in. foam insulation, a builder can brace a corner with 1/2-in. panels and cover them with 1/2-in. foam to bring the corner out to the depth of the rest of the insulated wall.*

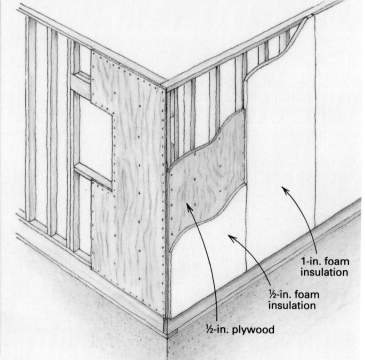

1-in. foam insulation

1/2-in. foam insulation

1/2-in. plywood

Drawings: Christopher Clapp

Shear walls

Here on the West Coast, many buildings require shear walls, and it takes an engineer to determine what combination of sheathing and metal connectors goes into a shear wall. Unlike sheathed-and-framed walls, a shear wall includes special measures to hold the house frame together at its weakest points—where the foundation meets the frame and where one floor meets another or the roof.

Seismic anchors—One common feature of a shear wall is that the framing is more-securely connected to the foundation. Most often this connection begins with anchor bolts. Common anchor bolts are L-shaped threaded rods embedded in the foundation that protrude up through the mudsill. When engineers design shear walls, they determine what type, what size and how many anchor bolts are necessary.

Some workers just jab anchor bolts into the wet concrete stemwall. Anchor bolts should be connected to the rebar (see p. 109 for more on installing anchor bolts).

In a shear wall, anchor bolts can be connected to the wall framing, not just to the mudsill. Seismic anchors, such as Simpson Strong Tie's HDA/HD series (photo above) are what engineers are calling for more and more to make this connection. These welded, heavy-gauge metal hold-downs prevent a building from pulling off its foundation. A seismic anchor fastens to an anchor bolt and is bolted to a post. Usually, the post is a 4x4 nailed in the wall like a stud. Although old-style seismic anchors had to be held up above the mudsill to reduce the chances of the bolt's splitting the post, the newer seismic anchors are redesigned to rest on the mudsill.

The building plans will indicate where a seismic anchor belongs and what kind to use. A shear wall usually has a seismic anchor at each end, whether the wall is 4 ft. long or 40 ft. long to resist uplift from both directions. Seismic anchors may also be required at corners and at each side of openings.

All about panels—I'm not an engineer, so I can't tell you everything there is to know about shear panels. Engineers have to consider not only what forces need to be resisted but also the number of openings in a wall, the size and grade of the lumber, the height of the building and many other factors. One thing I know from experience is that shear walls often require a better grade of sheathing than the stuff used on floors. Commonly, plans call for ⅜-in. thick structural I or II (the highest and next-highest grades) plywood on shear walls with studs at 16 in. o. c. and ½-in. structural I or II when the studs are 24 in. o. c. Check your plans and codes to be sure what grade of sheathing is called for.

A single 8-ft. panel normally will cover the framing from the bottom plate to the double top plate. On taller shear walls, I either nail a row of blocks at 8 ft. to provide a nailing surface for the end joints, or I use longer panels. On multistory houses, I like to nail 4x9 or 4x10 panels

Connecting foundation and frame. A seismic anchor is bolted to the foundation and to a 4x4 post; its job is to keep the building on the foundation. The seismic anchor is mounted above the mudsill to keep the through-bolts from splitting the post.

from the mudsill to the middle of the second-floor rim joist.

Panels can be installed horizontally, too. When plywood is nailed up horizontally, the majority of its veneers also run horizontally, making it stiffer under shear stress than vertical plywood. But to take advantage of that stiffness, the edges of horizontal plywood must be nailed to blocking. No matter how I lay up the panels on a shear wall, I always nail the edges to something—studs, rim joists, mudsill or blocking. If a panel doesn't break on the center of a stud, I either rip the panel to fit, or I nail an extra stud in the wall to provide adequate backing.

I try to place panels so that the seams fall somewhere in the middle of window and door openings rather than along an opening's edge. And I often sheathe right over doors and windows. Then, when the whole wall has been sheathed, I go back with a chainsaw or a reciprocating saw and cut out the openings. (see *FHB* #78, pp. 38-43 for more on using chainsaws.) There may be a required ⅛-in. gap for expansion between sheets. I achieve this gap by tacking a couple of 8d nails between sheets.

It's all in the nailing—An engineered plan includes the nailing schedule, which indicates the exact type of nail to use—its length, the size of its head and its shank diameter—as well as the spacing between the nails. Panels are nailed to hold-down posts and other framing members with a nailing schedule that can read, for example, 10d commons placed at 2-4-6. In other words, the nails in each panel are 2 in. o. c. along the perimeter of the building, 4 in. o. c. around edges that butt into other panels, and 6 in. o. c. in the inner part, or field, of the panel. That's a lot of nails!

Pneumatic nailers are great for nailing off shear walls. Just be careful to adjust the air pressure so that the nails are driven flush with the surface of the sheathing. If they break the skin on the panel, they're more likely to rip through the panel when the wall is stressed. Also, don't hold the nailer

Framing anchors. At this garage-door opening, metal strapping connects trimmer studs and headers; above it, three clips are nailed to the top plate and the rim joist to transfer shear forces.

directly in front of your face. If a nail hits a metal strap, such as a metal brace beneath the sheathing, it could drive the nailer straight back into your face.

Some building codes will not allow clip-head nails in shear walls, so make sure you're using full-head nails in your pneumatic nailer. And keep shear-wall nails back ⅜ in. from the panel's edge to reduce the chances of their splitting out.

Strapping—The sill can be attached to the rim (or band) joist and the rim joist to the bottom plate with framing anchors (bottom photo, above), such as Simpson's model A-35F. These flat metal clips are often nailed around the perimeter of the building every 16 in. o. c. Metal straps up to several feet long can make a positive connection between different floors. These straps are nailed into hold-down posts, rim joists and second-floor posts. Framing anchors add a lot of labor, but they'll keep your structure on the foundation and in one piece when the big one hits. —*L. H.*

Plumb and Line
Without this final step of straightening the walls, the care taken during framing will have little effect

by Don Dunkley

It's strange to think of a completed wall frame as being a kind of sculpture that needs final shaping. But that's just what it is. Until the walls have been braced straight and plumb, they can't be sheathed or fitted with joists or rafters without producing crooked hallways, bowed walls, ill-fitting doors and roller-coaster roofs.

The production name for getting the frame plumb, square and straight is *plumb and line.* The job doesn't take very long—three to six hours for most houses—but it's essential. After the frenzied pace of wall framing, plumb and line can be a welcome relief. It requires at least two carpenters (three's a luxury) working closely together. The work is exacting, but not hard, and there's a sense of casual celebration in having finished off the wall framing.

Stud-wall framing is based on things being parallel and repetitive (see the article on pp. 25-33). If you plumb up the end of a wall, then all the vertical members in the wall will be plumb in that direction. And if the bottom plate of the wall is nailed in a straight line to the floor, then getting the top of the wall parallel to the bottom is easy: just plumb the face of the wall at both ends, and make the top plates conform to a line between these two points.

Plumb and line is a fluid process in which walls are braced individually, but the sequence of operations is important. Although you can start at any outside corner, walls should be plumbed and then lined in either a clockwise or a counterclockwise order, since each correction will affect the next wall. Once the bottom plates have been fully nailed to the floor or bolted and pinned to the slab, the exterior walls are plumbed up and the let-in braces are nailed off, or temporary diagonal braces are installed to prevent the wall from racking (the movement in a wall that changes it from a rectangle to a parallelogram, throwing the vertical members out of plumb). Then line braces are nailed to the walls to push or pull them into line at the top and to hold them there. Last, the interior walls are plumbed and lined with shorter 2x4 braces.

Plumbing and lining can begin once all the intersecting walls are well nailed to their channels or corners. On each of these walls,

Holding wall intersections tight while the lapping double top plate is nailed off can make the difference between a plumb frame and having to make a lot of compromises later on.

he end stud has to line up perfectly with the channel flat or corner studs. It's also wise to make sure that the heights of the walls match up. If they don't, it's usually because the end stud isn't sitting down on the bottom plate. Also, all double top plates must be nailed off where they lap at corners and channels. Be sure that there aren't any gaps here (a common defect in hastily framed houses), and that the end studs and plates are sucked up tight (photo facing page).

Fixing the bottom plates—The next job is to toss all the scraps of wood off the slab or deck and sweep up. This way you'll be able to read the chalklines on the floor. If the frame is sitting on a wood subfloor, the bottom plates have likely been nailed off. On a slab, though, you'll have to begin by walking the perimeter of the building putting washers and nuts on all the foundation bolts. Tighten the nuts a few turns before using a small hand sledge to knock the plates into alignment with their layout snaplines. The slight compression created by the nut will keep the plate from bouncing when you hit it.

Getting the plates right on the layout is critical, because the bottom of the frame determines the final position of the top of the wall. Once the plates are where you want them, run the nuts down tight. The fastest tool for doing this is an impact wrench, found mostly on tracts and commercial jobs. Next best is a ratchet and socket, with last place going to the ubiquitous adjustable wrench.

Wherever bottom plates butt end-to-end on a slab, both pieces must be fastened down with a shot pin. There are many powder-actuated fastening devices currently available; the gun I like the best is the Hilti. Unlike many of the older guns, which have space-hogging spall shields, it is slim enough to get into tight spots between studs. The model that I've used (DX-350) is fitted with a multi-load magazine, so it can be reloaded in seconds. It also uses a pin and washer that are made as a single unit—a big advantage.

If there are two of you working, one person can set the pins into the top of the plate with a tap of a hammer. The other person can follow, slipping the gun barrel over the shaft of the pin and firing. The pins should be placed close to the inside edge of the plate so that you don't blow out the side of the slab. To prevent the bottom plate from splitting out at the ends, you can nail a ⅜-in. or ½-in. plywood scrap over it and then fire the pin through it. Since doorways take a lot of abuse, a pin should be shot into the sill at each side of the opening.

After all the outside walls are fastened down, you can line the bottom plates of the inside walls and shoot or nail them down. Be especially careful on long parallel runs like hallways—any deviation will really shout out when the drywall is hung and the baseboard installed. On interior partitions, one drive pin or nail every 32 in. should be sufficient. Make sure to hit the flanking bays of doorways, and both sides of bottom-plate breaks.

Setting up—Once all the bottom plates are secure, spread your bracing lumber neatly throughout the house. Interior walls will need 8-ft. and 10-ft. 2x4s. Exterior walls will need 14 and 16-footers for line braces, and shorter diagonal braces if the wall doesn't have let-ins. You'll have to guess how much bracing you'll need. More is always better than less, but figure that you'll need a line brace every 10 ft. and a diagonal brace on any wall over 8 ft. long without a let-in.

To plumb walls I use an 8-ft. level with two small aluminum spacer blocks that fit over the lip of the level near each end, and are tightened with handscrews. When the level is held vertically against a wall, the spacers butt the bottom plate and double top plate and prevent bowed studs from interfering. You don't need to buy an 8-footer; any level that is dead-on accurate will do if you extend it to the length you need (photo right). Select the straightest 2x4 around, cut it to 8 ft., and tack a short piece of 1x to one edge of the 2x4 at each end for a spacer. Then attach the level to the other edge of the 2x4 using duct tape or bent-over 8d nails.

Plumbing the walls—Exterior walls are plumbed from their corners. Choose any one as a starting point. Position the level at the end of the wall so that you are measuring the degree to which it's been racked. The bubble will indicate which way you have to rack the wall back to plumb. To do this, you can make a push stick out of a 9-ft. to 10-ft. 2x4 that is largely free of knots. A more flexible push stick can be made from two 1x4s nailed together face-to-face. Cut a slight bevel on the bottom end to keep it from slipping. To use the push stick, place the top of it in a bay about midway down the length of the wall to be racked, up where the stud butts the top plate. The stick should angle back in the opposite direction from which the top of the wall should be moved. Plant the bottom end on the slab or deck right next to the bottom plate. Pushing down on the center of the stick will flex it, exerting pressure on the top of the wall. Keep pushing and the wall will creak into a plumb position.

Racking walls takes coordinated effort. If you're on the smart end of the things (using the level), you need to shout out which way the wall must move and how far. Although overcompensation is usually a problem at first, soon your partner (who's on the push stick) will get used to what you mean by "just a touch" and be able to make the bubble straighten up as if he were looking at it.

Once you've racked the wall to plumb, hold it there permanently by nailing off the let-in bracing (photo right). But before you do that, have your partner keep the same tension on the push stick while you plumb the other end of the wall as a check. If it's plumb also, nail off the braces on the wall as fast as you can. Here it's handy to have a third person to do the nailing while you keep your eye on the bubble. If for any reason (poor plating or a gap where the top plates butt) the wall isn't

Plumbing. The length of the level you use is a lot less important than being sure that it is dead-on accurate. Here a 4-ft. level is attached to a straight stud with bent-over nails. Using 1x spacers at the top and bottom of the stud (or manufactured aluminum spacer blocks on an 8-ft. level) will ensure an accurate reading despite bowed wall studs.

Keeping it plumb. While the let-in brace that will keep the wall plumb is being nailed off, this short wall is being held plumb by tension on a push stick made of face-nailed 1x4s.

plumb on the opposite end, you'll have to split the difference.

Walls that are shorter than 8 ft., walls with lots of doors and windows, and exterior walls that will get shear panel, sheathing or finished exterior plywood won't have let-in bracing. For these walls, a temporary 2x4 brace must be nailed up flat to the inside of the wall at about 45°. Be sure that none of these braces extends above the top plate. Use two nails at the top plate, one nail each into the edges of at least three studs and two nails into the bottom plate. As with all plumb-and-line bracing, drive the nails home (you won't be pulling these braces until joists, rafters and sometimes sheathing are in place). Continue this process until all exterior walls are plumbed.

Lining—The next step is to get the tops of the exterior walls straight. If you've got a practiced eye, you can get the best line of sight on a wall by standing on a 6-ft. stepladder with your eye right down on the outside edge of the double top plate (photo above). This is the way I usually line, but for absolute accuracy you can't beat a dryline.

Set this string up by tacking a 1x4 flat to the outside edges of the top plates at each end of the wall. Then stretch nylon string from one end of the wall to the other across the faces of the 1x4s, and pull it tight (a twist knot looped and tightened on a nail will hold a taut line). The 1xs hold the line away from the wall so that if the wall bows, the string won't be affected. To determine whether the wall is in or out at any given point, use a third scrap of 1x as a gauge between the string and the top plates. If an inside wall intersects with the outside wall, I plumb the end of the interior wall, and then check the rest of the wall with the 1x gauge.

To make corrections in the wall and then hold it there, use line braces. These are 2x4s on edge that are nailed high on the wall and angle down to the floor or the base of an interior wall. To be most effective, they should be perpendicular to the plane of the wall. If a wall is at least 8 ft. long, it gets a brace—even if it's already straight. Braces not only correct the walls, but also make them secure enough to walk on while laying out and nailing joists or rafters, and keep these ceiling members from pushing the walls outward.

Usually, I begin lining a wall by sighting it quickly for bows. My partner nails braces where the deviations are worst. Then we brace off the rest of the wall where needed. If the wall is pretty straight, we scatter braces every 10 ft. or so, and do them in order.

The top of the line brace is attached to the wall first. Face-nail it to a stud just under the top plate with two 16d nails, making sure that it doesn't run beyond the exterior plane of the wall. If a header is in the way, face-nail a vertical 2x4 cleat to the header with three or four 16d nails, and then edge-nail the brace to the cleat. Line-brace connections are shown in the center and bottom drawing panels on the next page. Trimming an eyeballed 45° angle on the top of the brace will allow it to hug the wall and give you better nailing.

Setting the braces also requires good communication. While you pay attention to the string, your partner either pulls or pushes the brace according to your directions, and then anchors the bottom end when you say that it's looking good. You can usually find an inside wall to tie the bottom of the brace to. This won't affect plumbing the inside wall (which will be done later) as long as you nail the line brace to the bottom of a stud down where it butts the bottom plate. Use at least two 16d

Scissor levers for lining stubborn walls

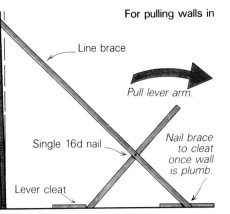

For pulling walls in

Line brace

Pull lever arm.

Single 16d nail

Nail brace to cleat once wall is plumb.

Lever cleat

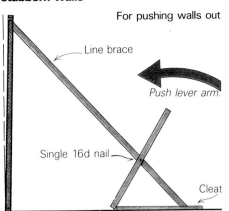

For pushing walls out

Line brace

Push lever arm.

Single 16d nail

Cleat

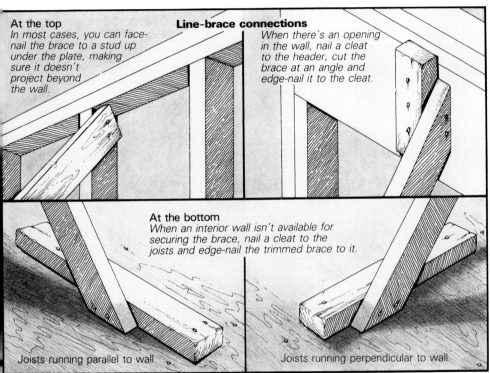

Line-brace connections

At the top
In most cases, you can face-nail the brace to a stud up under the plate, making sure it doesn't project beyond the wall.

When there's an opening in the wall, nail a cleat to the header, cut the brace at an angle and edge-nail it to the cleat.

At the bottom
When an interior wall isn't available for securing the brace, nail a cleat to the joists and edge-nail the trimmed brace to it.

Joists running parallel to wall

Joists running perpendicular to wall

Persuading stubborn walls. Using a 2x scrap (a 3-footer is ideal) for a kicker under a line brace will bring a wall in at the top. Straightening walls quickly requires the carpenter setting the line braces to make adjustments in or out a little bit at a time with the constant direction of his partner who is eyeballing the wall.

Lining. For absolute accuracy, there is no substitute for a dryline. But if you've got a good eye and an experienced partner you can get straight walls in a hurry (facing page). Either way, use the top exterior corner of the double top plate to gauge the straightness of the wall.

nails (or duplex nails) and drive them home. If an inside wall is not available, nail or pin a block to the floor, no closer to the outside wall than the wall's height. This will leave the brace at an angle of about 45°. Be sure to find a joist under the floor, since a floor cleat that is nailed only to the plywood will pull out. Trim the bottom of the brace in the same way you did the top, so that you can drive two good 16d nails through it into the cleat.

Most walls are relatively easy to move in or out, but every house has a notable exception or two. Walls full of headers can be a real pain to line because they are so rigid. Bringing the top of a wall back in is usually more difficult than pushing it out because you have to work exclusively on the inside of the wall. One effective technique is to use a *kicker* (photo above). Toenail the line brace flat (3½-in. dimension up) to the header and nail a block above it so it won't pull out under tension. Then toenail the bottom of the brace into a joist, and cut a 3-ft. 2x scrap. Toenail one end of it to the floor and wedge the top under the line brace so you have to beat on the kicker to get it perpendicular to the line brace. This bows the brace out, bringing the wall back in

(although occasionally it pulls the toenails loose on the line brace). When the wall lines up, end-nail the brace to the kicker.

You can also use a parallel interior wall to help pull in an exterior wall, or use a scissor lever that nails temporarily to the line brace. Scissor levers can be used to push or pull, depending on where the blocks are nailed (top drawing panel, above).

Line the exterior walls one at a time until you come full circle. Try not to block off entries and doorways with braces, but don't skimp either. I've learned to double the normal bracing on exterior walls that will carry rafters for a vaulted or cathedral ceiling. Also, if this ceiling includes a 6x ridge beam or a purlin 18 ft. or longer, I like to place the beam on the walls just after I've plumbed them. This way you don't have to pick your way through a forest of braces with that kind of load.

Interior walls—The last step is to plumb up the inside walls. Start at one end of the house and work your way through. You'll develop a sense of order as you go, and soon find that the remaining walls are already nearly plumb, and that the frame is beginning to become rig-

id and to act as a unit. Quite a few of the interior walls will have let-ins; if they don't, use the 8-ft. and 10-ft. 2x4s as diagonal braces on the inside of rooms. Where there are long hallways, I usually cut 2x4 spreaders the exact width of the hall measured at the bottom, and nail them between the top plates every 6 ft. to 8 ft. The spreaders will keep the entire hall a uniform width, and when the wall on one side is plumbed, the wall opposite will be too.

Once all the interior walls have been plumbed, go back to any long walls to check for straightness, and then throw in one more line brace for good measure. This is also the time to make a last check on exterior walls to make sure the interior-wall plumbing you did hasn't thrown a hump in the works. Then go around the entire job, shaking walls to make sure there's no movement. This kind of precaution means that when you're rolling joists or cutting rafters you won't have to measure the same span or run every few feet along a wall for fear of a bow, or worry that the walls have acquired a lean over the weekend. □

Don Dunkley is a framing contractor and carpenter in Cool, Calif.

Cutting Multiples

A few tricks can save time when cutting studs, blocks and cripples

by Larry Haun

One thing that I've noticed in the years I've been a carpenter and teacher is that cutting multiple pieces can be a terrible time-waster. If, for instance, you asked beginning carpenters to cut 100 blocks and didn't give them any further instruction, it would take them about 100 minutes to do the job. That's because most people tend to cut one block at a time. And it takes about a minute to find a piece of wood, measure it for length, mark it and then make the cut.

That's fine if you have one or two blocks to cut, but not if you have 100.

On most any job there are numerous occasions when carpenters will need piles of blocks, cripples, headers, trimmers or studs all cut to the same length. Some builders make a cutting list that contains the size, the length and the number of these items and then submit the list to their lumber companies along with the rest of the order. Large lumber companies often have gang

saws, and with the press of a button, a saw operator can make a pallet of blocks in no time. The blocks can be shipped to the job site with the rest of the order.

Most of us, though, don't operate that way. When we need a rack of cripples, we set up right in our work area and cut. There are several methods for cutting multiple pieces that are a lot faster than cutting them one at a time. The two keys are: cut more than one board at a time,

Pulling to a line. When gang-cutting 2x4s on edge, the author starts by pulling the stock to a straightedge, in this case a straight 2x4, with the claws of his hammer.

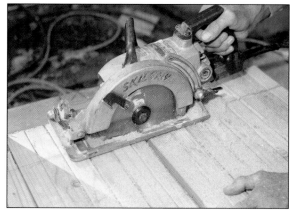

Measure once, cut twice. The author snaps a line between marks on the two outside 2x4s and then cuts on the line with the blade of his saw at full depth. After the first cut has been made, each 2x4 is picked up in turn, and the cut completed. The stock does not have to be marked again.

which is called gang-cutting, and don't measure each board individually.

Use a radial-arm saw and a stop block—
For many builders, the radial-arm saw is the preferred tool for gang-cutting wood to length. Some carpenters mount their radial-arm saw right on the back of a pickup so that the tool is readily available. I have mine mounted on its own trailer so that the saw can be pulled from job to job.

To cut multiple pieces with the radial-arm saw, I build a simple table out of a 14-ft. or 16-ft. long 2x12 (or two 2x6s) and nail a 2x4 fence to the backside. To make repetitive cuts to the same length, all you have to do is attach a stop block to the table at the correct distance from the blade (bottom photo, p. 64). With a stop block in place, you can feed in several 2xs at a time on edge and turn out a pile of blocks or cripples in short order because you're not stopping to measure and mark each piece. If a radial-arm saw isn't available, you can do the same thing with a power miter saw. The stop may be screwed, nailed or clamped to the worktable. If you have many cuts to make, be sure the stop block is well secured so that it isn't gradually forced out of position when you push the end of a 2x against it.

It pays to watch for sawdust that gets trapped between the stop block and the 2x you are cutting because your cuts can be thrown off by it. You can avoid the problem by cutting a chamfer, or bevel, on the bottom edge of the stop block so that sawdust is pushed out of the way.

Cut right on the lumber pile—
For many builders the circular saw may be the only saw available on the job site, and it can be used to cut multiple pieces right on the lumber pile. Usually, 2x stock is delivered in bundles or lifts with pieces lying flat; the material can be cut before it's taken off the pile. Before cutting the pieces to length, you need to flush up one end so that all the blocks or studs will be the same length. This can be done easily by holding the edge of a straight stud against one end of the top layer and pushing all the 2xs to a straight line. If the material is too heavy to push, you can stick the claws of your hammer into them and pull the 2xs against the straightedge. Once the 2xs are even at one end, you can measure down the two outside pieces to the point where the cut should be made and snap a line across the pile. Then set the saw to cut 1½ in. deep and make the cut (top photo, p. 65).

Rack up blocks to cut all at once—
Once a job is underway, there often is scrap material around that you will want to use up. Let's say you need 60 cripples, each cut 3 ft. long. Gather up pieces of 2x scrap and line them up, on edge, against a bottom sill or another straightedge on the job (top photo, facing page). Measure up 3 ft. from the flush end and snap a chalkline across the entire line of 2xs. You can cut the full depth of the blade along the line (bottom left photo, facing page) and then pick up each piece individually to finish the cut (bottom right photo, facing page). You won't have to mark the studs a second time. If you only have a few to cut, it is easier to lay them flat on the floor and make the cut in one pass.

This technique also works well when making a lot of short blocks (photo below). If the scrap pieces of 2x are long enough, you can snap a number of lines across them at the right spacing

Making small blocks. If the blocks are short, many of them may be made from a small stack of 2x scrap. The author snaps a series of lines across the scrap and makes the cuts, keeping the blade to the same side of the line each time so that the blocks are the same length.

and then make a number of cuts with your saw without further measuring or marking. Just remember to keep your blade on the same side of the chalkline each time so that the blocks are the same length.

Lay out multiples with a framing square—
You can also use your framing square to speed the cutting of blocks. I frequently need a number of lap or eaves blocks to fit between joists or rafters. When scrap pieces are not available, it is easier to lay out and cut the blocks from full-length stock right on the deck than it is trying to muscle 2x10s or 2x12s onto a saw table. If you are cutting 14½-in. blocks (the length that fits between rafters or joists 16 in. o. c.), align the 14½-in. mark on the inside of the blade of the square with the end of the 2x. Then draw a line across the 2x using the inside of the tongue of the square as your guide. Now move the 14½-in. mark to the line and repeat the process (bottom photo, facing page). Using this method, you can work your way down a 2x quickly. When it's time to cut the blocks, hold the sawblade to the same side of the line each time to ensure that each block is the same length.

Get a bigger saw—A beam saw can cut through a 2x6 on edge and can also be used to make the ridge cut when gang-cutting common

Big saw, big cut.
A large-capacity beam saw is capable of cutting 2x4s or 2x6s on edge. The author just lines up a stack of 2xs, snaps a line and makes the cut. No second cut is needed. A 2x4 spacer prevents the saw from cutting the plywood floor.

Cutting blocks on the radial-arm saw. Using a radial-arm saw is a fast way to cut multiple pieces on a job site. Several 2xs can be cut at the same time, and an adjustable stop block makes it unnecessary to measure stock for each cut. A bevel on the bottom edge of the wooden stop block helps prevent a buildup of sawdust, which will throw off the accuracy of the cut.

rafters. A beam saw is just an oversized circular saw; my Makita has a 16-in. blade. So if I'm cutting blocks from 2x4 or 2x6 stock on edge, the entire cut can be made in a single pass. In this case, I place a flat 2x under the stock near the chalkline to hold the 2x material away from the deck (top photo, facing page). A little paraffin on the blade makes the cut go easier.

If you don't have a beam saw and don't want to invest in one, a chainsaw attachment will in-crease the capacity of your circular saw. One kind is the Prazi Beam Cutter (Prazi USA, 118 Long Pond Road, Unit G, Plymouth, Mass. 02360; 800-262-0211), which the manufacturer says can be put on and taken off in a few minutes. The company makes models that fit either a worm-drive circular saw or a sidewinder. Another brand is the Linear Link VCS-12 (Muskegon Power Tools, 2357 Whitehall Road, North Muskegon, Mich. 49445; 800-635-5465). The Linear Link can be purchased either as a complete saw or as a kit to convert your worm-drive, although it's not sold as a quick-change accessory. Both the Linear Link and the Prazi increase the cutting capacity of your saw to 1 ft. at 90°. This type of saw is especially useful when gang-cutting the ridge cut on wide rafters. □

Larry Haun is a carpenter in Los Angeles, Calif. Photos by Larry Hammerness.

Why move the pile? Studs, blocks and cripples also can be cut right on the lumber pile. The author starts by flushing up the ends of the top layer of 2xs. He snaps a line across the pile at the right length, sets his saw to 1½ in. deep and makes the cut.

Skip the tape. When cutting short lengths of blocking from a long 2x, there's no need to measure and square each block separately. Instead, the author uses his framing square to mark off blocks quickly. He aligns the end of the 2x with the correct measurement on the inside of the square's blade, uses the tongue to mark a square line across the 2x and then moves the square along the 2x to repeat the process.

Don't Forget the Blocks

Blocking makes a house more stable and makes it easier to install drywall, plumbing and siding

by Larry Haun

Along with a mortgage, homeowners sometimes get an unwelcome riddle or two not long after they move in to their new house. Why, for instance, did the handrail on the staircase loosen up so quickly? And what caused the towel bar in the bathroom to fall off? These are telltale signs that something was forgotten when the house was framed: Backing and blocking that should have been included as the house was built almost certainly were overlooked by the framing crew. Blocking aids the installation of everything from bathtubs and handrails to siding and wainscoting, and it en-sures that things will stay put. And forgotten blocks can be more than an inconvenience. Some blocking used in construction is important structurally, such as edge blocks used under floor sheathing. Still other blocks, fire stops for instance, are safety features. The trick is to know when blocking is needed and where to nail it. ☐

Larry Haun is a carpenter who lives in Los Angeles, Calif. His book, The Very Efficient Carpenter, *and video were released by The Taunton Press in 1992.*

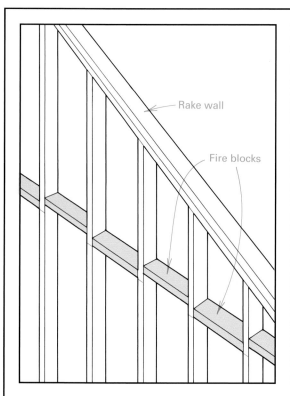

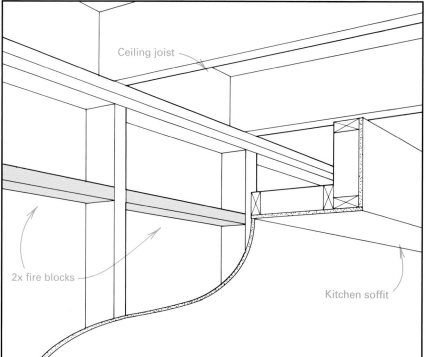

Blocks for safety—Fire stops (or blocks) were important in balloon-framed walls, when uninterrupted stud cavities rose two stories instead of just one. When fires started, those spaces acted just like flues, creating drafts that spread flames quickly. With the introduction of platform framing, fire blocks in standard-height walls became history. But many of to-day's codes do require fire blocks in walls that are more than 10 ft. high, like rake walls (left drawing, above).

When they are needed, cut the blocks to length (14½ in. for studs 16 in. o. c.; 22½ in. when studs are 24 in. o. c.) and nail them between the studs while framing the wall on the floor. One 16d common nail on one end and two nails on the other end provide plenty of support. Snap a chalkline on the wall and alternately nail the blocks above and below the line. There are two good reasons to stagger the blocks like this: It makes it easier to nail them, and a plaster wall won't crack when applied over them. Cracks can develop in plaster if the blocks are nailed in a straight line.

Fire blocks are also used to close off one framed area from another. A dropped ceiling in a hallway, for example, or a kitchen soffit should be closed off from the framed walls by a row of fire blocks (right drawing, above). If a fire did get started in one of these places, blocks would help contain it. Codes in my area call for 2x fire blocks, which provide more protection than 1x material.

If you do use fire blocks, be careful not to nail them in at head height. Someone may get hurt as they go from room to room through a stud wall. There's nothing like having your eyes crossed by walking into a fire block that catches you squarely in the forehead.

Drawings: Vince Babak

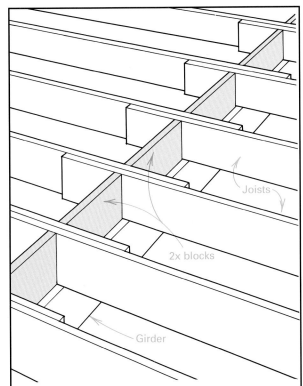

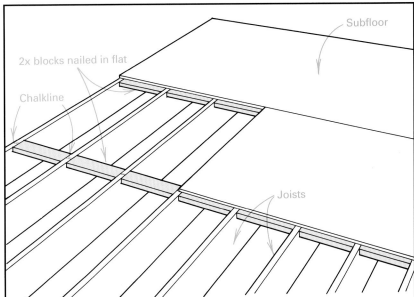

Joist blocks—For greater structural stability, joists should be blocked where they lap over a girder or a bearing wall (drawing above). Some codes require joists to be blocked over all bearing points, not just where joists overlap. Midspan blocks are no longer required by most codes, but see what is required in your area. Blocks at the lap should be 13 in. long when joists are 16 in. o. c. You can put the blocks in as you install the joists so that you won't have to toenail both ends in later.

Edge blocks—If your subfloor is straight-edged sheathing, you may be required to use edge blocking to support the subfloor (drawing above). These 2x blocks, which are nailed flat between joists, support panel edges but generally are not needed when T&G subfloor is used. Snap a chalkline every 4 ft. across the joists and nail in a row of blocks centered on this line, two 16d nails at each end.

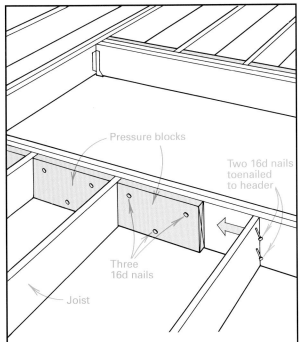

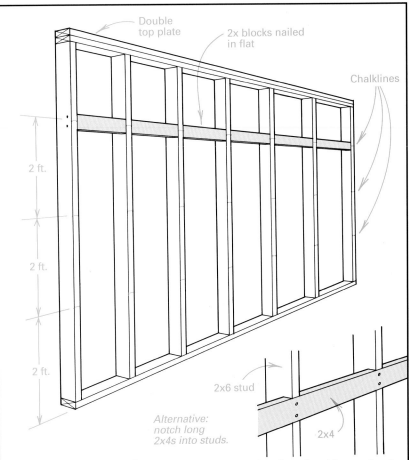

Pressure blocks—Ceiling joists that run into headers around an opening, such as an attic access or a skylight, often are supported by joist hangers. But if the joists are not going to carry much weight, you can use what I call pressure blocks (drawing above). These 2x blocks are cut to size (14½ in. for joists 16 in. o. c.) and are nailed between every joist.

Blocks for siding—To support vertical siding, such as board and batten, I nail rows of blocks about every 2 ft. o. c. in exterior walls (drawing above). Just snap chalklines while framing the wall and nail the blocks in flat, flush with the exterior side, so that they won't be in the way of insulation. When the exterior walls have 2x6 studs, it's just as easy to notch in long 2x4s for siding support. This is fairly easy to do when the wall is flat on the floor during the framing process. Straighten up the wall, and snap two lines the width of a 2x4 apart every 2 ft. across the studs. Then make a saw cut 1½ in. deep on these lines and cut out the notch. A 2x4 nailed in across the studs supports the siding and straightens bowed studs.

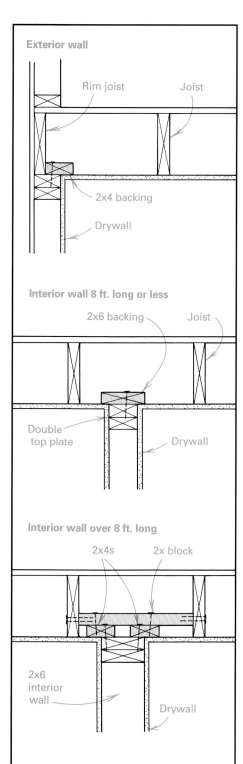

Exterior wall

Rim joist Joist

2x4 backing

Drywall

Interior wall 8 ft. long or less

2x6 backing Joist

Double top plate

Drywall

Interior wall over 8 ft. long

2x4s 2x block

2x6 interior wall

Drywall

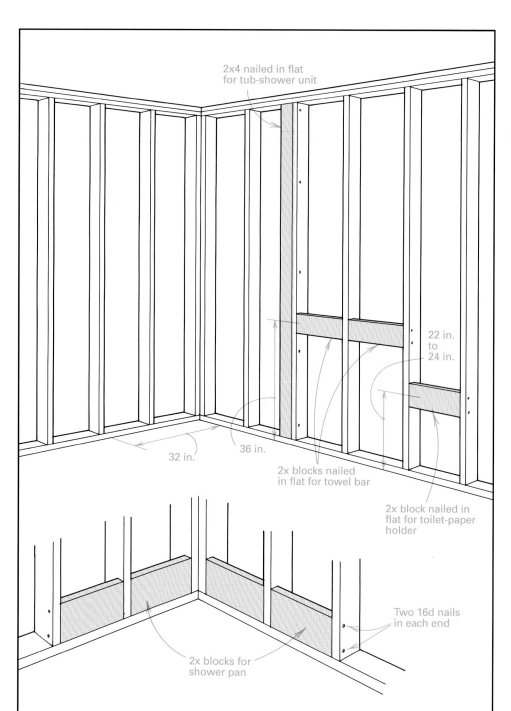

2x4 nailed in flat for tub-shower unit

22 in. to 24 in.

32 in.

36 in.

2x blocks nailed in flat for towel bar

2x block nailed in flat for toilet-paper holder

Two 16d nails in each end

2x blocks for shower pan

Backing—Backing is the term used to describe the 2xs nailed to the top of walls to provide nailing for drywall. When joisting for a second floor, or any ceiling that will be covered with drywall, backing has to be nailed on the top plates of walls that run parallel to the joists. On an outside wall, nail a 2x4 flat on the top plate and let it hang over the plate line into the room below (top drawing). On short inside walls, a 2x6 or two 2x4s side by side can be nailed flat on the double top plate for backing on both sides of the wall (middle drawing). On interior walls over 8 ft. long, a block nailed between the joists and into the backing will help hold the wall straight (bottom drawing).

Bathroom backing and blocking—Pay special attention when laying out and framing bathrooms. Extra backing is needed for most fiberglass tub-shower units or for tubs with tile above. A standard size for tubs is 30 in. wide, so when I detail the plates for stud locations, I measure out from the inside corner 32-in. and mark for a stud (drawing above). This stud is nailed outside of the 32-in. mark. Then back toward the inside of the 32-in. line, I mark the location of a flat stud that will nail right against the first. This flat stud will give the plumber a place to nail the flange on the tub-shower unit.

Showers that will be tiled need blocking at the base so that a waterproof pan can be attached. Here is a good place to use up scraps of 2x10 or 2x12 rafter and joist stock. These blocks sit on the bottom plate; nail them in flat between the studs around the perimeter of the shower.

Anyone who has ever tried to attach towel bars to studs in a bathroom knows how frustrating this is. People usually wind up using Molly or toggle bolts that hold everything in place—until someone uses the bars for calisthenics. These bars are often held 36 in. from the floor. So flat blocks nailed between studs at the proper height and location offer solid wood to which bath fixtures can be secured. Scraps of 2x6 or larger are better than 2x4s because you don't have to be as precise in locating them during framing.

I center toilet-paper holders at 22 in. to 24 in. from the floor. If the holder will be surface-mounted, nail a 2x block flat between the studs. If the holder will be recessed, the block is held to the backside of the wall. Finish carpenters feel good all day when they cut a hole in the drywall and find a block inside to which they can screw a recessed toilet-paper holder.

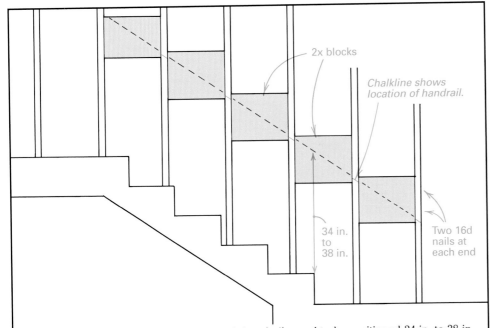

Stair-rail blocks—My codes state that stair handrails need to be positioned 34 in. to 38 in. plumb off the nose of the tread. Rather than try to guess the exact points at which a finish carpenter will attach the rail, I snap a chalkline at the proper rail height of the stairs. Then I pick up some joist or rafter scraps and cut blocks to fit between every stud space (drawing above).

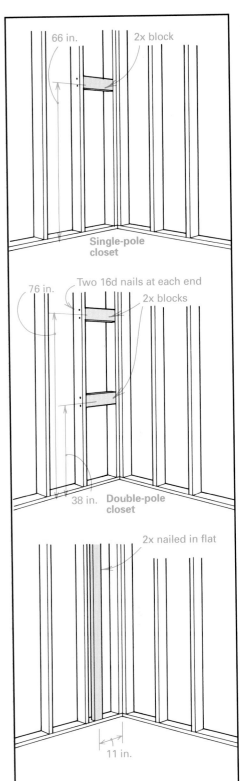

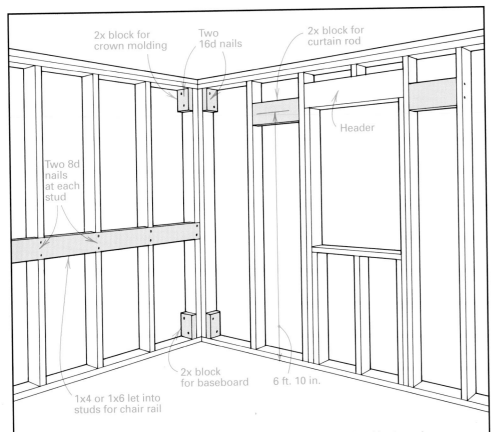

Special blocking—When installing wide baseboard and crown molding, it's nice to have scraps of 2x4 nailed at the top and bottom of the corner studs and at the bottom of king studs on door openings (drawing above). Blocks nailed flat between studs at the proper height offer solid backing for wainscoting, chair rail and picture rail, too. It may be just as fast to let in a continuous 1x4 or 1x6 in these areas.

Curtain-rod holders usually can be attached to the king studs or trimmers. But sometimes the homeowners will use curtain rods that extend beyond those studs. If this is the case, a 2x nailed flat at the window ends, and centered at 6 ft. 10 in. from the floor (the standard window height), will give adequate backing. Another place for special blocks is in the kitchen for a cabinet that will hang over an island or peninsula. The 2x blocks should be nailed between joists.

Closet blocks—The height of closet shelves and poles can vary. Around here a single pole is 66 in. off the floor. In a double-pole closet, the poles go in at 38 in. and 76 in. To provide solid backing for the poles, nail 2x blocks flat between the studs at the desired height on closet sidewalls (drawing above). It's also a good idea to provide backing for shelf cleats or adjustable shelf hardware that will be attached to a closet wall. Standard shelving in closets is 11½ in. wide. Rather than try to figure out closet arrangements during framing, I like to center a stud flat on the sidewall 11 in. out from the back wall.

Simple Curved Corners

Bending plywood and joint compound make graceful curves in white walls

by Scott M. Carpenter

When my brother told me he wanted to convert his basement into a playroom for his children, I winced. I knew he needed the help of a professional builder and that he assumed I would render that help for the relatively modest payment of lunches and dinners (and the periodic use of his services as a master electrician).

He was looking for ideas, and I had two good ones. My first good idea was to let my brother and his wife clean out the basement. My next good idea had to do with the framing. If this space was to be a playroom, how could we make it more playful and less dangerous? Two words: curved corners (photo right).

Considering the options—Tight-radius curves (2 in. or 3 in.) are easy to make if you use a drywall product called Gypcove (distributed by Pioneer Materials, Inc., 9304 East 39th St. N, Wichita, Kan. 67226; 316-636-4343), but the larger-radius curves I had in mind for the walls and the window returns required something else. In the past, I framed curved corners with radius-cut plywood top and bottom plates that were ribbed with studs and skinned with two layers of ¼-in. drywall. This drywall was wetted down to make it pliable, yet it still didn't bend smoothly; it pleated and took time to finish.

Although I didn't want to hang around at my brother's longer than necessary, and despite the fact that my brother is, well, a cheapskate, we liked the idea of curved corners. Luckily, a material called bending plywood offered a solution that saved enough on construction time to make it more than worth the extra cost of materials.

Bending plywood—I discovered bending plywood at my lumberyard, but it's more commonly obtained from suppliers of cabinet-making materials (see sidebar facing page). Several species and thicknesses are available; we chose ⅜-in. three-ply mahogany. It cost me $28 per 4x8 sheet, but that was several years ago, and I was able to buy directly from a distributor. What makes bending plywood different from standard plywood is that it has only three plies, and the inner ply is much thinner than the outer ones. This makes bending plywood flexible. The thinner middle ply has its grain running perpendicular to the outer layers, which gives the plywood its strength. Some sheets are laminated with the grain running in the 4-ft. direction, others in

Kid-friendly. **Curved corners add comfort and safety to a room for children. Bending plywood creates these corners, which are spackled, taped, skim-coated and painted exactly like drywall.**

the 8-ft. direction. The direction of the grain determines the direction the sheet will bend. As an experiment, I bent an 8-ft. sheet in a 6-in. radius. The creaking noise it made warned me to run for cover, but it didn't break.

I calculated that a single sheet would yield two 8-ft. high corners with a 9-in. radius. We had three corners and four window returns, so we bought two 4x8 sheets (grain oriented in the 8-ft. direction) and one 8x4 sheet (grain oriented in the 4-ft. direction) for window-return corners.

Framing the curves—We began work on the basement by marking the layout of the walls on the floor. We drew all outside corners square,

then we measured 9 in. in from each corner: the radius of each curved corner. When we framed, the straight walls stopped at the 9-in. marks.

The studs at the 9-in. marks were wood 2x4s; the rest were metal studs. I use metal studs in basements partly because the quality of lumber is going downhill, and decent wood studs are hard to get, but mostly I use metal studs because they're easy to work with, especially for furring out block walls. A metal stud wall made of 1⅝-in. studs has many advantages over a wood stud wall, including the fact that it's not affected by basement moisture.

With each wall stopped 9 in. short of the square corner, I felt there should be something in the

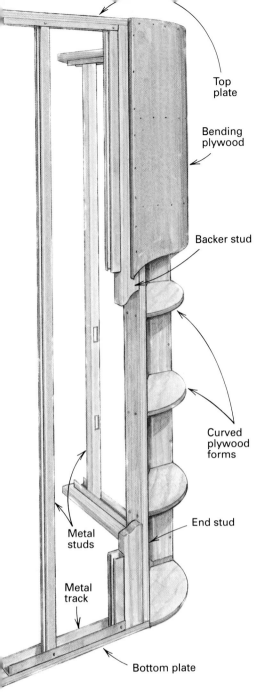

Top plate

Bending plywood

Backer stud

Curved plywood forms

End stud

Metal studs

Metal track

Bottom plate

Curved wall corner
Top and bottom plates stop 9 in. from the square corner; curved plywood forms fill the open corner. Alongside each end stud, a backer stud was placed on the flat, and the bending plywood was screwed to it.

open corners to wrap the bending plywood around. So I made curved forms. I cut 9-in. radius circles from ¾-in. plywood, quartered the circles and trimmed the points at 45°. I nailed these curved forms into the open corners like shelves at 16 in. o. c. (drawing above), nailing through the end studs and into the forms.

Getting the bends—An additional stud was turned sideways and nailed to each end stud, on the straight side of the wall, to serve as a backer for the drywall. Lapping the bending plywood halfway over this backer stud made taping easier. Through trial and error we found that if the butt joint between the bending plywood and the dry-

wall is made 3 in. away from the start of the radius, it is easier to feather the bulge of the tape line into the curve of the corner. Turning the backer stud sideways in the wall gave us room to lap the bending plywood 3 in. onto the wall framing with plenty of space left to fasten the drywall.

We also found it easier to cut the bending plywood after it was installed. By leaving it wide, we could use the extra width for leverage as we bent the plywood around a corner. Our system was to screw one edge of a full sheet to a backer stud, and then my brother bent the plywood around the corner while I screwed it to the curved forms every 2 in. with 1¼-in. drywall screws and finally to the adjacent stud. Then we marked and cut the bending plywood to width in place with a circular saw and finished the cut with a reciprocating saw. The job became physically more difficult when we were installing the last piece from each full sheet, which was only about 27 in. wide.

Having a moisture content between 6% and 10%, bending plywood takes tape and joint compound just as drywall does. The straight walls were skinned with ½-in. drywall that butts the ⅜-in. bending plywood. We feathered the thickness variance with joint compound, taped the seam, and then skim coated, or covered the entire corner with a layer of joint compound, to smooth out the bending plywood's rough texture. To check the transition from straight wall to curve, I examined the wall with a halogen lamp, a light that cruelly reveals imperfections.

Curved window returns—Typically in basement renovations the exterior walls are furred out to accommodate insulation and electrical wiring, and the windows end up set deep in the wall. We wanted to avoid this look, so we furred out the walls to create 9-in. radius window returns. These window returns turned out to be successful because they reflect more light into the room than square returns, making the basement brighter.

The fact that my brother decided to replace the old cellar windows with thermopane awning windows, which crank open at the bottom, gave us an opportunity to simplify the window returns. Before mortaring the new windows in place, we built out their frames with two layers of plywood glued on the flat (drawing right). The first layer is recessed back from the inside edge of the window frame; the outside layer of plywood is flush with the frame, making a pocket all the way around the window.

After installing the modified windows at the correct height, we slipped the drywall ceiling into the top pocket of each window frame. In the side pockets we squeezed a bead of construction adhesive and inserted bending plywood and screwed it to the wall studs (photo above). The tension created by bending the plywood forces it into the pocket and holds it firmly. No other backer or plywood bending forms were required. A drywall sill set into the bottom pocket completed the window return.   □

Scott M. Carpenter is a builder in St. Louis Park, Minn. Photos by the author.

Window-return corners

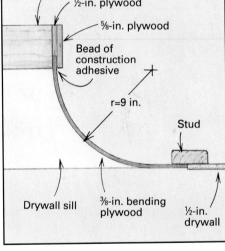

Brightening the basement. Bending plywood opens up window returns in furred-out walls. The window frame was built up with two layers of conventional plywood, creating pockets that hold the drywall sill and ceiling and the bending plywood returns.

Window frame

½-in. plywood

⅝-in. plywood

Bead of construction adhesive

r=9 in.

Stud

Drywall sill

⅜-in. bending plywood

½-in. drywall

Finding bending plywood

Bending plywood, also known as wacky wood and wiggle wood, may be tough to track down in some areas. Check the Yellow Pages under "Plywood and Veneers," or you can call either Danville Plywood Corp. (P. O. Box 2249, Danville, Va. 24541; 804-793-4626) or North American Plywood (10309 Norwalk Blvd., Santa Fe Springs, Calif. 90670; 310-941-7575) for the names of local distributors.
    —Rich Ziegner, assistant editor at Fine Homebuilding.

Metal Connectors for Wood Framing

Galvanized steel folded into sturdy, time-saving devices

by Bruce R. Berg

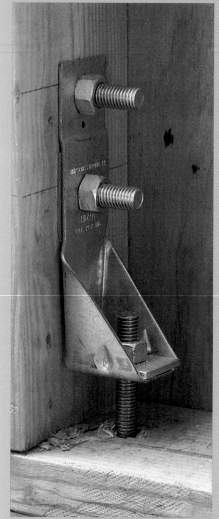

Hold-downs are tension ties between the foundation and the framing. The distance of the bolts from the ends of the studs must be at least seven times the diameter of the bolt. Often engineers will specify greater distances.

A swallowtail scarf joint is an elegant way to make a tension tie between adjoining beams, and a housed dovetail will anchor a floor joist to a girder for the life of a structure. But not many construction budgets have an allowance for the extensive and meticulous cutting and fitting that it takes to achieve these time-honored joints. These days, most structural connections in wood-frame buildings are made with steel connectors because they are affordable and easy to install. Also, their structural values have been carefully tested and documented. Consequently, steel connectors are widely accepted by codes and building officials.

The companies that make metal connectors (see the sidebar on p. 77 for sources of supply) offer their products in a remarkable number of configurations. Their catalogs include not only illustrations of the connectors, but also tables that list specifications such as the dimensions of the lumber and the appropriate connectors, their design loads, and the number and size of nails it takes to achieve that rating.

Joist hangers are probably the most common type of metal connector on a construction site, but if you need them you can get connectors to anchor a scissors truss to a bearing wall, adjustable post bases or metal clips that allow you to install outdoor decking without visible nails. This article takes a look at the principal types of steel connectors. Within these categories there are many variations that you can use to solve specific construction problems.

Concrete-to-wood connectors—If you have ever struggled to lift a framed wall onto a protruding row of anchor bolts, you are familiar with a potential source of frustration. Despite everyone's best intentions, the holes in the sill plate sometimes don't line up with the bolts, and the plate has to be redrilled. Or a stud lands on an anchor bolt, requiring a nasty-looking notch in the bottom of the stud.

One alternative to anchor bolts that circumvents these problems is the MAS galvanized steel anchor from Simpson (drawing A, facing page). It resembles a Y with a ladle-like cup on the bottom leg that gets embedded in concrete. The branches of the Y are wrapped around the mudsill or up the side of a stud and secured with nails. Prior to the pour, these anchors can be positioned by tacking them to the formwork. And because they emerge from the concrete at the edge of the footing, you don't have to hand-trowel around a bunch of anchor bolts.

Another sheet-metal anchor from Simpson (called the MA) can be attached to the mudsill before you pour your foundations (drawing B). The anchor's pointy, arrowhead shape allows it to slip easily into the screeded wet concrete.

Seismic anchors (sometimes called hold-downs or tie-downs) are frequently specified by architects and engineers when part of a structure needs lateral bracing and there is only a narrow wall section in which to provide it. The narrow wall is stiffened with plywood for shear

Post anchor

Post anchors are used to attach wood columns to concrete.

D

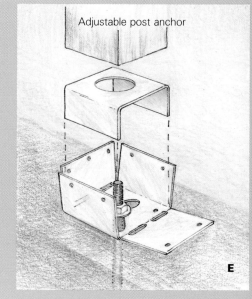

Adjustable post anchor

E

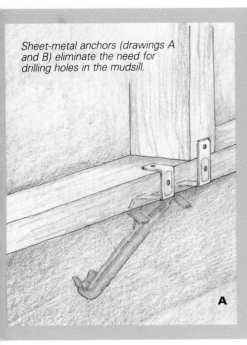

Sheet-metal anchors (drawings A and B) eliminate the need for drilling holes in the mudsill.

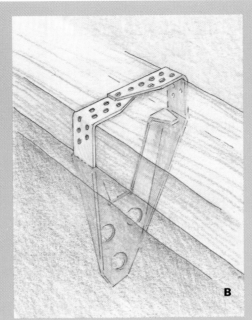

A

B

The simplest hold-downs are steel straps with a deformed end that is embedded in concrete. Installed horizontally, they can anchor joists or purlins.

C

strength, but when a horizontal load is applied to the top of a stiff panel, it wants to lift away from one of its corners. A hold-down provides resistance to this uplift.

Hold-downs come in two basic varieties. The first is a deformed strap that is set into the wet concrete at the location of a post or stud (drawing C). It is then nailed or bolted to the post, sometimes with as many as 24 nails.

The other variety uses a foundation bolt that rises through the plate and through the bottom of a gussetted, welded heavy angle to which it is fastened (photo facing page). The angle is then bolted to the stud or post at the perimeter of the plywood panel. Because of the seismic activity here on the West Coast, builders frequently use pairs of these hold-downs linked by a threaded rod to create a tension tie between two floors.

Very similar to the deformed-strap hold-down are purlin and joist anchors. They are embedded horizontally into the concrete or masonry wall,

aligned with the top of each framing level. When nailed off to the joists or purlins, they allow the horizontal diaphragm to work together with the wall structure, and prevent the walls from leaning outward.

A peek under many backyard decks will reveal row upon row of concrete pier blocks supporting a forest of 4x4 posts. To distribute their loads evenly, the blocks are typically set into a bed of wet concrete. A chunk of pressure-treated pine or redwood is attached to the top of the block for toenailed connections. Pier blocks are easy to work with, but they have their limitations. The wood blocks sometimes detach from the concrete, and they are so small that they often split when you drive nails into them. Also, they provide little resistance to lateral or uplift loads, and can't be replaced if required. A far better way to anchor posts, such as those used to support decks, is the post or column base.

Post anchors are made in several configura-

tions—each for a particular application. One kind is designed to be placed into wet concrete (drawing D, facing page), and its positioning must be precise. Another type uses an anchor bolt to secure it to concrete, and it has a slotted adjustment plate (drawing E) that allows you to tinker with its alignment in the event of a slightly misplaced bolt. You can also use this type of post base on a cured concrete slab or footing by tying it down with an expansion bolt, concrete nails or powder-actuated fasteners.

Some post anchors have a standoff plate that elevates the post about 1 in. above the concrete, which prolongs the life of the post in moist areas. Under really wet conditions, you can use an elevated post anchor (drawing F) to get the wood several inches above the concrete.

Cleveland Steel makes a dressy post base out of cast aluminum that lends a sturdy appearance to a column, and elevates the wood above the concrete (drawing G). The base is secured with

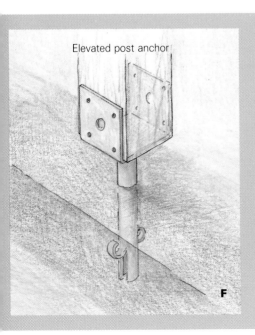

Elevated post anchor

F

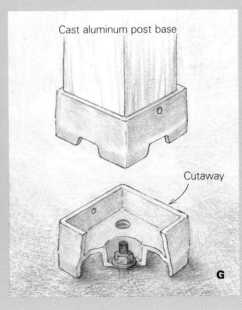

Cast aluminum post base

Cutaway

G

Heavy-duty post bases are excellent foundations for fences. The steel pad under the post is raised slightly above the concrete to keep the post dry. Grout between the steel pad and the concrete ensures good bearing.

a single anchor bolt, and weep holes allow an escape route for rainwater.

Since most post bases allow little room for adjustment once they've been installed, you've got to be scrupulously accurate as you embed them in the wet concrete. Use a string line or a transit to align rows of post bases as they are set in the concrete. Lumber-crayon marks on the formwork can ensure the accuracy of your spacing in the other direction.

Sometimes builders use post bases to support beams, as on a low deck. In this situation you can first prepare the forms, then place the beam in its final position on top of falsework with the post bases attached to the beam. Pour your concrete, take out the falsework and your beam is ready to carry its load without the aid of a wooden column.

Post bases also come in heavy-duty versions. These are made of thicker steel than the standard bases, and have longer straps that extend up the sides of the post. One good application of the heavy-duty post base is to anchor fence

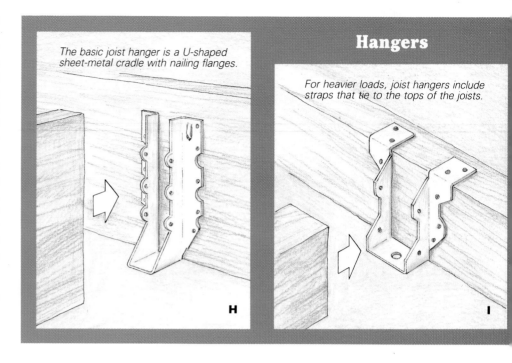

The basic joist hanger is a U-shaped sheet-metal cradle with nailing flanges.

Hangers

For heavier loads, joist hangers include straps that tie to the tops of the joists.

H

I

Straps and bracing

Steel straps are useful tension ties throughout wood-frame buildings. Here a steel strap is used to resist the uplift load on a beam.

posts (photo previous page). Excavate your post hole and use a short section of Sonotube to bring the level of the concrete about 6 in. above grade. As you fill the hole with concrete, insert a #4 rebar in the center of the hole before you embed the heavy-duty post base. Bolted to this kind of a base, your fence posts won't rot off at grade in ten to twenty years.

Hangers—The basic hanger is a galvanized strip of 14-ga. to 18-ga. steel folded into a U-shape (drawing H). Hangers are made in various sizes to accommodate typical framing lumber. At the bottom of the U, the metal widens to form a seat for the joist or purlin. Properly installed, the member rests snugly against the seat of the hanger. Flanges on the legs of the U turn outward, and nails driven through holes in the flanges secure the hanger to the beam; other holes allow fastening to the joist.

Representatives of the companies that make

steel connectors say that the most common mistake made in installing their products is inadequate nailing. Because the shear strength of the nails is the limiting factor in the strength of the connection, it's critical that you install the connectors with the nails specified by the manufacturer. Commonly available wire nails in the correct diameter are often longer than the thickness of the framing member, so you have to use "joist-hanger nails," which are simply short versions of common nails. Also, nails that are used in outdoor locations or driven into treated wood should be galvanized.

Another typical mistake is to cut the joists too short. The gap between the beam and the end of the joist should be no more than 3/16 in., and closer is better.

When you are installing joist hangers on a beam that is deeper than the joists, you have to snap a chalkline on the beam to align the bottoms of the hangers. Because joist material

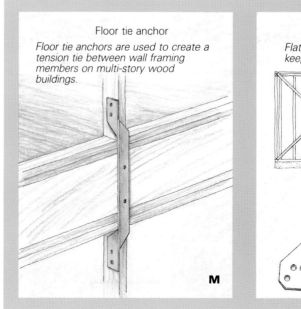

Floor tie anchor

Floor tie anchors are used to create a tension tie between wall framing members on multi-story wood buildings.

M

Wall bracing

Flat straps applied in a crossing pattern keep a framed wall from racking.

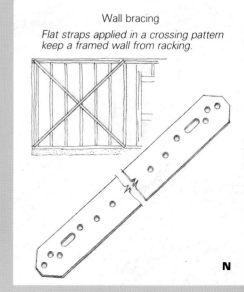

N

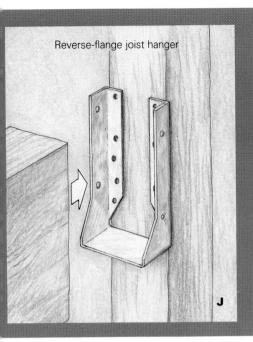

Reverse-flange joist hanger

J

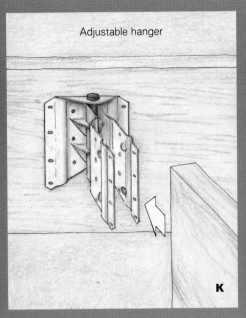

Adjustable hanger

K

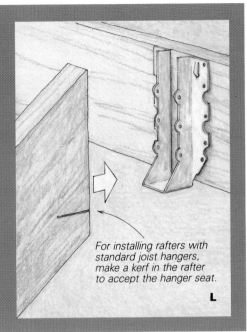

For installing rafters with standard joist hangers, make a kerf in the rafter to accept the hanger seat.

L

sometimes varies slightly in depth, position your hangers to accommodate the deepest joists in the lumber pile. Then add shims to the hanger seats to bring the shallower joists flush with the top of the beam.

To hang joists that carry heavier than normal loads, use hangers with a top flange (drawing I). When you nail down the subfloor, don't worry about the thickness of the metal draped over the joists—a few hammer blows will compress the wood fibers enough to make the floor lie flat.

On most hangers the flanges turn outward, but you can also get them with reverse flanges (drawing J). Reverse-flange hangers can be useful in tight spots, such as a window retrofit when you have to add a header and there isn't room for trimmers to carry its load, or when two perpendicular beams meet at one post.

Not all hangers are designed to carry members perpendicular to a beam. Most companies make 45° hangers as standard items, and some

manufacturers, such as Panel Clip and Cleveland Steel, are set up to make "specials" to fit unusual framing needs. They can make hangers that are sloped to carry rafters, skewed at angles other than 45°, or a combination of the two. Simpson makes a hinged hanger that will skew to 60° and slope up to 30° (drawing K).

Another way to handle a rafter with a conventional hanger is to cut a kerf in the end of the rafter that is the same depth as the hanger seat and perpendicular to the plumb cut (drawing L). The hanger seat tucks into the kerf.

Straps and bracing—Sometimes called tension ties, steel straps are applied to wood-frame construction in numerous ways. For example, if the plumber cuts through the top plate in a wall, a steel strap can restore the structural integrity of the plate without adding much bulk to the framing. Other common uses of the steel strap are to tie opposing rafters together across a

ridge, and to resist the uplift of a cantilevered beam (photo facing page).

One of the simplest yet most useful connectors I have used is the twist strap. A typical one is made of 16-ga., 1¼-in. wide galvanized steel bent so that the faces of the two ends are 90° to each other. They are often used in pairs on framing members that cross one another to prevent the wood from twisting. I've also used them as hangers to suspend old ceiling joists from new beams during remodel work.

Another version of the twist strap is the floor tie anchor (drawing M). A 90° degree twist at each end of this strap allows it to be attached to the framing on multi-story buildings to make a tension tie between floors. It does the same thing as a pair of hold-downs, for less money.

Metal wall bracing (drawing N) is another form of strap tie that can speed things up on a job site. While not nearly as strong as a well-nailed plywood diaphragm, it is at least as strong as 1x4 let-in bracing, and you don't have to cut notches to install it. To prevent racking, the flat variety needs to be applied in pairs to form an X or a V.

Whenever I'm remodeling older homes that have diagonal blocking, I enhance the shear strength of the wall by adding a flat wall brace to the line of blocks. I nail it securely to each block and each stud. If I'm working on an interior wall, I use a power plane to cut a shallow groove for the brace across the studs so it won't telegraph through the finished wall.

Another type of steel wall bracing comes in a T or L section (drawing O). It is let into a kerf cut across the studs in a straight line, providing the framed wall with both compression and tension bracing in a single strip.

Metal bridging (drawing P) is so much faster to install than wood bridging that it is an ideal example of how steel connectors have added to the efficiency of wood-frame construction. Prongs on the ends of the bridging eliminate the need for nails. You drive one end into a joist about 1 in. from its top edge, then the other end

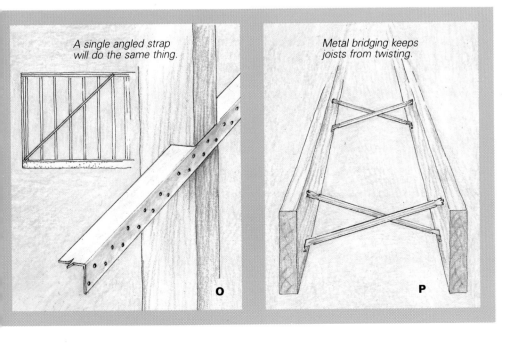

A single angled strap will do the same thing.

O

Metal bridging keeps joists from twisting.

P

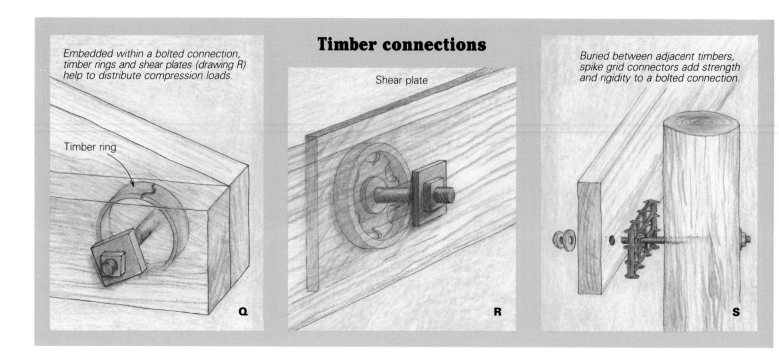

Timber connections

Embedded within a bolted connection, timber rings and shear plates (drawing R) help to distribute compression loads.

Timber ring

Shear plate

R

Buried between adjacent timbers, spike grid connectors add strength and rigidity to a bolted connection.

Q

S

into the adjacent joist 1 in. from its bottom edge. Because they require no nails, they don't develop nail squeaks. But don't let the pairs of bridging touch one another or they will make noise. Keep them 1 in. apart, and remember that on spans over 16 ft., you need to use two sets of bridging to conform to most building codes.

Timber connections—Timbers are typically bolted together, and because they are usually supporting substantial loads, a lot of pressure is concentrated on the bolted connections. Timber rings, also called split rings, and shear plates are two metal connectors that are used in concert with bolts to spread out compression loads and shear forces, reducing the potential for crushed wood fibers around the bolts.

Timber rings (drawing Q) are steel rings that ride in matching grooves cut into adjacent timbers. A grooving tool that resembles a hole saw is used to cut the grooves for the rings, and at

the same time it bores a hole for the bolt that runs through the center of the rings. When installed, the rings are hidden from view, captured by the pair of timbers.

Shear plates (drawing R) are similar to timber rings, but they are used to make metal-to-timber and concrete-to-timber connections. Both Cleveland Steel and TECO are suppliers of timber rings, shear plates and the grooving tools necessary to install them.

TECO also makes spike grid connectors, which are used to add strength and rigidity to the joints between heavy timbers. Resembling medieval instruments of torture (drawing S), the grids consist of rows of spikes protruding from a malleable iron matrix. They fit between two timbers at a bolted connection, and a threaded compression tool is used to apply enough pressure to the timbers to embed the spikes. For securing a pole to a timber, TECO makes a spike grid that is curved on one side and flat on the other.

Truss clips—If you install nonbearing partitions in a building with trusses overhead, you must not create a rigid connection between the truss and the top plate of the wall. The bottom chord of a truss moves up and down as the loads on it change, and if you don't take its vertical movement into account, the truss can become overloaded.

Steel angles called truss clips can be used to attach a partition wall to a truss while still allowing the truss to move (drawing T, facing page). A slot in the vertical leg of the angle accepts a nail into the lower chord of the truss, anchoring the top of the partition while allowing the chord to move up and down.

Cleveland Steel makes a connector plate, shown in drawing U, that uses the same slot principle to anchor a scissors truss to a wall plate. In this application, the truss wants to move in a horizontal direction, and the slots allow a full inch of movement.

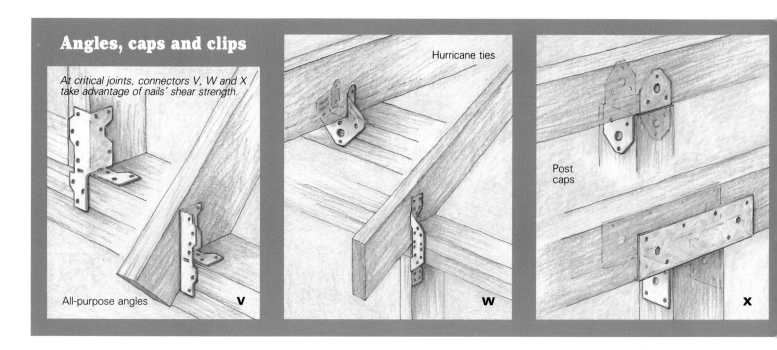

Angles, caps and clips

At critical joints, connectors V, W and X take advantage of nails' shear strength.

Hurricane ties

Post caps

All-purpose angles   **V**

W

X

Angles, caps and clips—All the companies that manufacture steel connectors make multi-purpose devices that are known as angles, angle clips or reinforcing angles. Their function is to connect butt-joined framing members without toenailing. The more elaborate versions are partially slotted at the fold and have bend lines that allow them to fit a variety of intersections (drawing V, facing page).

Hurricane ties and seismic anchors are another way to avoid toenailing at critical connections. They are folded to wrap around rafters and top plates (drawing W), where they are secured with nails that are working in shear.

Post caps too are designed to get the nails or bolts to work in shear where a post and its beam come together (drawing X). Most are made of 16-ga. galvanized steel, but heavy-duty versions made of 3-ga. painted steel are a standard item from Simpson. Post caps resemble post bases, and in fact some are made to accept a piece of rebar so they can be partially embedded in concrete to become post bases.

By using plywood clips (drawing Y) you can avoid having to use blocking under all the edges of the plywood. The clips will keep the edges of the plywood from seriously deflecting under heavy loads. I use two plywood clips for 16-in. rafter spacing and three for 24-in. rafter spacing. But I don't use them for hot-mopped roofs, because the potential for a little deflection is still there. Instead I'll block under all edges or I'll use T&G plywood.

A slick clip that lets you build a deck without exposed fasteners is made by Philips Manufacturing. It is a galvanized steel angle with prongs on one side that grab the decking on its edge (drawing Z), eliminating rusty nailheads and indented moons from misplaced hammer blows. And since a portion of the clip is sandwiched between adjacent pieces of decking, the clip also acts as a spacer to ensure good drainage.

Specials—Sometimes no commercially available connector can solve a given problem. If I can find something close to what I need in one of my catalogs, I give the manufacturer a call and ask about modifications. Frequently the company is more than willing to customize a connector for me, and the price has not seemed out of line. (Of course, be sure to ask what the delivery time will be.) To ensure accurate results, supply your fabricator with a full-scale drawing of the special connector that specifies the material to be used, all dimensions, angles and nail or bolt placement.

If you need a truly unusual steel connector, I recommend going to a local sheet-metal or welding shop. A good drawing of the connector is essential. If you are unsure about the loads that the custom connector may have to carry or resist, see if the welding shop can figure them out. Failing that, seek out an architect or an engineer for assistance. □

Bruce Berg is a construction supervisor at Christopherson & Graff, Architects, in Berkeley, Calif.

Truss clips

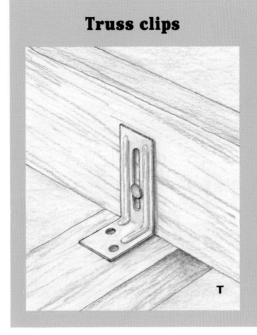

T

Scissors-truss connectors allow a truss to move relative to the framing while remaining anchored.

U

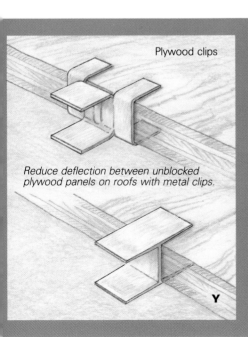

Plywood clips

Reduce deflection between unblocked plywood panels on roofs with metal clips.

Y

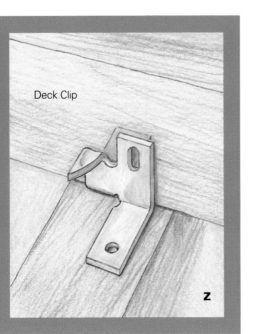

Deck Clip

Z

Sources of supply

Cleveland Steel Specialty Co.
14400 South Industrial Ave.
Cleveland, Ohio 44137
(800) 251-8351

Dec-Klip
Philips Manufacturing
460 2nd St.
Lebanon, Ore. 97355
(800) 544-0124

Harlen Metal Products
300 West Carob St.
Compton, Calif. 90220
(213) 774-8383

Heckmann Building Products Inc.
4015 W. Carroll Ave.
Chicago, Ill. 60624-1899
(800) 621-4140

Panel Clip
4203 Shoreline Drive
Earth City, Mo. 63045
(800) 521-9335

Silver Metal Products Inc.
2150 Kitty Hawk Rd.
Livermore, Calif. 94550-9611
(415) 449-4100

Simpson
1450 Doolittle Dr.
P.O. Box 1568
San Leandro, Calif. 94577
(415) 562-7775

TECO
5530 Wisconsin Ave.
Chevy Chase, Md. 20815
(800) 638-8989

Remodeling With Metal Studs

Skyrocketing lumber costs make steel studs an enticing, easily installed alternative to wood

by Tom O'Brien

Tools for steel framing should look familiar. All the tools you'll need for framing with steel are probably already in your toolbox. They include spring clamps, a 2-ft. level, a chalkline, a plumb bob, tin snips and a cordless drill. A screw gun (background) and locking C-clamp pliers are also helpful.

I recently went into a lumberyard to pick up a few 2x4s. The bill left me wondering if maybe I should have stopped off at the bank for a second mortgage. Sadly, anyone who has purchased lumber lately has probably had a similar experience. But what alternative is there to the high price and dubious quality of framing lumber? For more than ten years metal studs have been the answer for me. In Virginia, lightweight 25-ga. metal studs for remodeling are a little more than half the cost of good quality 2x4s.

I don't suggest abandoning wood completely in favor of steel. Despite cost advantages, metal framing does have drawbacks that limit its effectiveness for total residential framing. Load-bearing partitions require a more costly, heavier gauge steel that has to be cut with special tools and must be welded or fastened with expensive drill-tipped screws. Also, the thermal conductivity of steel makes insulating a steel-studded wall more difficult. For these reasons I still choose wood for framing exterior and load-bearing walls. But for other framing applications, I find light-gauge metal faster, cheaper and easier to work with than wood.

Steel framing is stable and uniform—Until lumber prices went out of sight, material costs for wood and steel were roughly the same. I was using metal framing then because my labor costs were lower. The reasons are simple. Metal framing is a manufactured product, which means that it's stable, straight and uniform. These qualities translate into time saved that would be wasted digging through piles of wood-framing stock, looking for acceptable material, and sorting and crowning at the job site. Product stability also eliminates the need to repair or replace metal-framing members that warp or distort after they have been installed.

Metal framing is easier to handle than wood because it weighs significantly less. Studs, for instance, come in easy-to-carry, interlocking bundles of ten (top left photo, facing page). Steel boasts other advantages over wood, including resistance to damage by fire, insects and weather. Steel framing is stocked in standard sizes from 1⅝-in. to 6-in. widths up to 20-ft. lengths.

Lightweight steel framing requires no special tools—Chances are most of the tools needed for metal framing are already in your toolbox.

Try carrying ten 2x4s like this. Steel framing weighs much less than wood, and studs come in interlocking bundles of ten, which makes them easier to carry.

Cutting lightweight steel is quick and easy with tin snips. After cutting through the flanges of the track, or stud, bend the waste end back and cut through the web with a slightly circular inward motion.

Clamping pliers aid assembly. Locking pliers with a C-clamp head keep the steel studs in place until screws are driven. Each stud gets one screw through each side, top and bottom.

These tools include a measuring tape, square, plumb bob, level (a magnetic level is handy but not necessary), chalkline and tin snips (left photo, facing page).

For most jobs, all that's needed to cut metal studs is ordinary, straight-cutting tin snips. Metal-cutting chopsaws that can slice through entire bundles of studs at one time are available for about $200. However, I've gotten along just fine all these years using tin snips to cut steel framing. Here's how I do it.

First, I mark the stud the same way I would a 2x4 except that I use a felt-tip marking pen (or a grease pencil) because it shows up better on metal than pencil does. I cut through the flanges on both sides of the stud with my tin snips, then turn the stud over and bend the cut end back. I then cut through the back of the stud (the web) with a slightly circular inward motion (photo top right). If you are unable to cut all the way through from one direction, just turn the stud around and finish the cut from the other side.

A VSR screw gun facilitates assembly—For fastening metal framing as well as attaching drywall, the one substantial tool purchase I would recommend is a variable-speed reversing (VSR)

screw gun. This tool streamlines the assembly process. I prefer a 0-2,500 rpm model because it is geared lower for more power and better control, but a 0-4,000 rpm model works fine and is usually less expensive.

Another tool that makes framing with metal much easier is 4-in. locking C-clamp pliers made by Vise-Grip. A pair of these is well worth the small investment. I also keep a couple of 2-in. spring clamps handy for suspending my plumb bob from the ceiling-track flange.

Self-drilling screws connect steel framing members—Screws are the fastest and easiest way to fasten metal framing. Framing members are usually joined with 7/16-in. type-S self-drilling pan-head screws, commonly referred to as framing screws. Avoid type-S12 screws with drill-point tips, which are designed for fastening heavy-gauge steel. These screws are more expensive and tend to strip out light-gauge metal.

To join framing members, I first clamp the two pieces in place with my Vise-Grips (photo bottom right). Next I put a screw on the magnetic bit in my screw gun and hold it against the track, keeping the gun as perpendicular to the track as possible. I start the gun as if drilling a hole, and

when the screw begins to penetrate, I back off slightly on speed and pressure and let the screw thread itself home. I advise using only a professional-quality #2 magnetic Phillips bit and changing the bit at the first sign of wear.

Drywall is attached to metal studs with type-S drywall screws, which have finer threads and sharper points than the type-W screws used with lumber. I use 1⅛-in. screws for single-layer drywall and 1⅝-in. screws where two layers are specified. I don't know anyone who still uses nails to install drywall, so hanging it on metal should not be a difficult transition; just remember to use a lighter touch on the screw gun.

Casing and baseboard can be nailed to wood blocking installed in the metal framing, but it is simpler to attach trim with finish screws, which look and work the same as finish nails. They have small, self-countersinking heads, which are designed to be driven below the surface of the wood and covered with putty. These days, most finish screws have square-drive heads that work better than Phillips heads but still require a bit of care to avoid strip outs. Again, these screw tips should be changed at the first sign of wear. The lengths of trim screws most commonly used are 1⅝ in. and 2¼ in. The shorter ones are easier to

The drywall is an integral part of the corner. When building a corner of steel, the drywall on the inside is attached to the last stud of the bywall before the last stud of the adjoining wall is installed (right). That stud is fastened with screws driven through the drywall (below). This method simplifies construction and eliminates the need for extra framing in the corner.

work with but are just barely long enough to apply ¾-in. trim over ½-in. drywall. I try to keep some of the longer screws handy at all times.

It is possible to attach softwood trim to a metal-framed wall without predrilling, but you're probably asking for trouble. If you don't predrill, screws may bend, creating large, ugly holes as they go in, or they may refuse to countersink. It's also difficult to keep the wood tight to the wall and to keep it from splitting. I usually keep a ⅛-in. bit chucked in my cordless drill and alternate between that and my screw gun, although I have drilled and fastened trim using only my cordless drill with a quick-change bit system. When predrilling, it's necessary to stop before the bit hits steel framing. The screw should be driven so that the head is countersunk about ⅛ in.

Basic framing process is similar to wood—
Except for a few simple differences, framing with metal is basically the same as stick framing with wood. Plates in steel framing are made of U-shaped channel, and the studs have a C-shaped profile (right photo, p. 78).

To frame a basic metal-stud wall, I start by laying out the bottom plate as I would with any wall. After establishing the two end points of the wall, I snap a line on the floor and mark on which side of the line I want the wall. I transfer this line to the ceiling using a plumb bob or a level alongside a steel stud and always check to make sure the top plate is marked on the same side of the line as the bottom. Next, I cut the top and bottom plates

out of track stock, screw them into place and mark the stud layout on the tracks. At this point I take a few stud height measurements at various locations. If they differ by ½ in. or less, I subtract ⅛ in. from the shortest measurement and cut all studs to that length. Because the studs fit between the flanges of the track and are screwed in place, it is perfectly acceptable for stud height to vary by as much as ½ in.

After cutting the studs to length, I stand them up, setting the bottom of each stud inside the track lengthwise and tilting the top into its approximate position in the top track. Next, I twist the stud a quarter-turn so that the flanges of the track grip it and hold it in place.

All of the studs can be erected in this manner before they are fastened. After orienting the studs on the layout marks, I roughly split the gap in stud length between the top and bottom track and clamp the bottoms. The bottom of each stud gets fastened with a screw driven through each side of the track (photo bottom right, p. 79). Finally, I climb a ladder and secure the tops.

A couple of steel-framing quirks need to be pointed out. First, most metal studs come with prepunched holes to accommodate plumbing and electrical systems. These holes need to be aligned before the studs are cut. Some of these punch-outs also have a definite top and bottom, which means that all the studs must be measured and cut from the same end.

Second, the open part of the steel studs should face the beginning of the stud layout. This place-

ment lets the drywall contractors know where to start their sheets. More important, it prevents the studs from distorting when the drywall is attached, which keeps the seams between the sheets flat and even.

No extra studs are needed in the corners— Framing with metal studs eliminates the need for special corner construction or extra framing because the drywall itself ties the intersecting walls together. When framing the corners of steel-framed walls, I first decide which wall will run by and which will butt just as in wood framing. I cut the track pieces for the bywall to the exact length and put them in position. Next, I measure and cut the track for the butt wall, leaving about a ¾-in. gap between the end of the track and the edge of the bywall track. This space is for the drywall. After cutting and positioning all the track, I double-check that each wall is laid out the same at the top and at the bottom and then anchor the track in place.

I lay out the studs for both the bywall and the butt wall, allowing for just a single stud at the end of each wall. Next, I cut and fasten all of the studs, except for the end stud of the butt wall. I leave this butt-wall stud floating until the drywall has been installed.

After the blocking is in place and plumbing and electrical lines are roughed in, the walls are ready for drywall. I start inside the room on a bywall and run the board through the corner to the end of the wall (top photo, facing page). Then I

hold the floating corner stud from the adjacent butt wall against the drywall I just installed and fasten it with screws driven through the back of the drywall and into the web of the floating stud (bottom photo, facing page). The top and bottom of the floating stud now can be screwed into the stud's track. This procedure ties the corner together and uses fewer studs than wood framing. The corner is completed structurally when drywall is applied to the outside of the walls.

Framing the intersection of two steel-studded walls is essentially the same as framing a corner. Again, no special framing is required. The track for the intersecting wall is kept short, and the end stud for that wall is back-screwed on after the drywall is installed on the main wall. Sometimes it's not possible to get behind the bywall to back-screw the intersecting stud. In these cases, I slide the last stud of the intersecting wall into position against the drywall and shoot pairs of 1⅛-in. drywall screws through the inside corners of the stud into the drywall at opposite 45° angles. A pair of these screws every foot or so holds the stud in place nicely until the drywall is installed.

To give a metal-studded wall additional rigidity, I stagger the seams of the drywall so that inside and outside sheets don't break on the same stud. Taping and finishing the drywall is the same as always: messy and tedious.

Door framing is lined with wood for nailing— Framing for doors in a steel-studded wall requires extra effort. First, the studs must be at least 2½ in. wide to accommodate a door jamb. I start framing by locating the position of the door in the layout on the floor. I mark the centerline of the rough opening, and I measure half the rough-opening dimension in each direction.

I stop the track for the bottom wall plates at the edges of the rough door opening. However, I install the studs 1½ in. back from each side of the rough opening, which allows room to line the rough opening with 2x stock. This wood makes it easier to attach the door jamb and casing. I always check the door studs for plumb before screwing them in.

To complete the door frame, I measure the rough-opening height off the high side of the floor, then add 1½ in. again for blocking. I mark this point on the stud with my marker and com-

bination square, then use a level to determine the height of the other side. I cut the door header from a piece of track that's exactly 11 in. longer than the rough opening, and I square lines across the header 4 in. in from each end. This cut leaves me with the rough-opening width plus 3 in. for the blocking on both sides between my marks. I then make 45° angle cuts through the flanges of the track at both lines.

After making these cuts, I bend the 4-in. flaps down and install the header with the flanges facing up (photo p. 81). I clamp the tabs to the door frame at the correct height and screw through the tabs into the studs with two framing screws on each side. Finally, I continue my stud layout on the header and install the cripples between the header and the ceiling track.

I complete the door frame by attaching 2x blocking to the inside of the door-frame studs (photo left). When I use 3⅝-in. studs, full-width 2x4s fit perfectly. But if I'm using 2½-in. studs to save floor space, I rip 2x6s down to about 2⅜ in. for the blocking. The blocking is then cut to length and attached to the steel studs with 1⅛-in. screws.

Steel framing is the best option for hiding masonry walls—Here in Virginia, I do a lot of renovations of old brick row houses. Concealing unsightly brick or masonry walls has always presented a problem. These wall surfaces are rarely smooth enough to accept drywall directly, and using furring strips usually requires a lot of fussing

Wood and metal work side by side. To create a proper rough opening for a door, 2x stock is added beside metal framing. Wood makes attaching door jambs and casings easier.

Drywall scraps help stiffen the wall. Pieces of drywall screwed to the studs with their edges against the wall behind help solidify a steel-studded wall that's only 1⅝ in. thick. The excess drywall is trimmed off later. Wood blocking is installed between studs to accept kitchen cabinets. The flanges of the steel studs have been flattened where the blocking passes by to keep the blocking flush with the stud face.

and shimming. Framing with 2x4s held away from the wall is fine but costly in space as well as money, and 2x2 walls seldom remain straight. My solution to this renovation problem is to use 1⅝-in. metal framing, which stays straight and requires little space.

First, I lay out the bottom plate 2⅛ in. off the brick (1⅝ in. for the stud width plus an extra ½ in. to allow for variations in the wall surface). After snapping the chalkline, I check a few points along the line to make sure the wall can be plumbed without hitting the brick. I also adjust the line for squareness to the room to accommodate any cabinetry, floor tiles or intersecting walls. If necessary, the chalkline can be moved farther out.

When I'm satisfied with my layout, I use a plumb bob to transfer the layout to the ceiling for the top track. Next, I install both top and bottom track pieces. If I'm securing the track to wood floors or ceiling joists, 1⅛-in. drywall screws work fine. But if I'm attaching the track to a concrete floor, I use 1-in. powder-actuated fasteners. I finish framing the wall by cutting and installing the studs.

Compared with a 2x4 wall, an uncovered metal stud wall may seem downright flimsy. However, drywall applied correctly to both sides stiffens the wall. When I'm framing along a masonry wall, I can't apply drywall to the back of the studs. Instead, I install small scraps of drywall to act as stiffeners (bottom photo, facing page). I hold each drywall scrap against the brick and

screw it to the web of the stud, taking care not to bow the stud out into the room. I leave the edges of the scraps long and trim them off after they have been installed.

If kitchen cabinets are being installed, it's necessary to add solid blocking to a metal-framed wall (bottom photo, facing page). This process can be done by first cutting up scrap 2x or ¾-in. plywood into 15¾-in. or 23¾-in. lengths, depending on the stud layout. Next, I mark all the pieces 1 in. from one end and cut a ¼-in. deep saw kerf on each side of the line at that depth. This double kerf slips over the lip of the stud flange and allows the blocking to be installed flush with the outside of the studs. I often use another method of installing blocking without making saw kerfs. For that method, I bend the stud flange flat where the 2x passes by. In either case the blocking is attached with two 1⅛-in. drywall screws driven through the flange of the stud on one end of the block and through the inside of the adjacent stud on the other end.

After all the blocking is installed, the plumbing and electric have to be roughed in before the wall can be closed. I try to hire subcontractors who are familiar with metal framing because some procedures differ slightly from wood framing. Plumbers must be sure to isolate copper pipe from the steel framing with plastic bushings or tape to prevent galvanic corrosion. Electrical boxes may be screwed to the sides of the studs, fastened to wood blocking or attached with clips made specially for metal framing. Many electri-

cians choose to run conduit through metal framing, but it's unnecessary as long as bushings are used to protect the wire sheathing from being damaged by the edges of the studs.

Before drywalling the outside of the walls, I sometimes insulate to provide thermal protection or soundproofing. Because metal studs are C-shaped and hollow, conventional insulation ends up being 1½ in. too narrow. Full-width insulation, 16 in. or 24 in., is available from your metal-stud supplier.

Framing soffits with steel is a breeze—When building interior soffits, I use metal framing almost exclusively. Whether it's a ceiling above kitchen cabinets, an enclosure for mechanical components or a transition between rooms with ceilings of different heights, I can build a soffit straighter and more efficiently with 1⅝-in. metal framing than with wood (photo below). Plus I don't need the arms of George Foreman to hold a section of lightweight steel framing over my head while driving screws.

For soffits, I use essentially the same framing procedure as I do with wood, although I prefer to build my soffits after the drywall has been applied to the adjacent walls and ceilings. This process makes layout easier, makes the soffit stronger and eliminates the need for blocking (nailers) at the corners. □

_____

Tom O'Brien is a carpenter and remodeler in Richmond, Virginia. Photos by Roe A. Osborn.

Old technology meets new. A soffit is created of steel framing to conceal ductwork in the basement of an old house. The vertical members that will hold drywall are attached to sandblasted beams more than a century old.

Header Tricks for Remodelers
Creative responses to unusual specs

by Roger Gwinnup

Between the Scylla of the homeowner's desires and the Charybdis of the house's structural needs, the course of the remodeler is often narrowly charted. When the request is to "take out that wall, but hide the header so it looks like nothing ever happened," the possibilities are usually at least two, but hopefully not more than two million. Given the vagaries of everyone who has ever worked on the house previously, the only way to proceed is to get out the old recipro-saw/scalpel and make an incision. Having done so several times, with instructions to "hide that header," I'd like to share some techniques that have worked for myself and my former partner Bill Pappas, now in Minnesota.

Heading off truss ends — I once built a small, gable-roof addition to a house that had pre-engineered roof trusses. The header had to run perpendicular, and adjacent to, the truss ends. I plumb cut the truss ends flush with the outside of the old framing (being careful not to disturb the truss plates) and removed enough plywood sheathing to allow access for nailing (drawing 1). Using a metal-cutting blade in my reciprocating saw, I cut the nails holding the truss to the wall and slid a joist hanger on the end of each truss. After installing posts at either end of the opening, I set the header and nailed the joist hangers to it. All that remained was to secure additional joist hangers to the other side of the new ceiling joists, remove the wall, and have a donut.

One advantage to this method, besides the concealed header, is that the existing wall supports everything until the new header is completely installed. Another advantage is that starting wall removal from the outside of a house leaves everything in place as a barrier between a habitable room and a construction zone. Depending on the situation, several wall components may be reusable, especially the insulation and the studs, if the nails have been cut from the top and bottom plates and the studs carefully twisted away from the drywall.

Balloon-frame header addition — I once worked on a two-story balloon-frame house with 2x6 walls where I needed to add a one-story addition with no visible header over the opening between them.

I put in a temporary wall inside to support the second floor; I then notched out the existing wall studs to receive the new header

(drawing 2, facing page). Because the studs in a balloon frame extend in one continuous length from sill to roof, I notched the header 1¾ in. into the wall studs to better support the existing second floor. For this particular job I used two Micro=Lams to form the header. The advantage of this is that I can go to my supplier with the spans and load factors, and they will calculate the size I need. I secured the first Micro=Lam to the framing with 16d cement-coated sinkers, then attached the second to the first with more 16ds and Max Bond Adhesive (H. B. Fuller Co., Building Products Div., 315 S. Hicks Rd., Palatine, Il. 60067; 708-358-9555), though any good construction adhesive will do. I installed support posts at each end of the opening, and used twisted hanger straps to connect each existing joist to each stud and to the Micro=Lam. Then I removed the studs below the joists and hung the new ceiling joists.

Concealed header in an attic — I once removed part of a bearing wall in the middle of

a house that supported the ceiling joists from each side of the house. In this case I went up in the attic with some 2x10s and laid them across the top of all the joists that were going to lose bottom support.

Then I installed twisted metal straps (drawing 3, facing page) to connect the header and the joists, went back downstairs, and removed some drywall and a stud from each end of the wall to be removed. After making sure that the posts themselves were supported below to the foundation, I slipped them in place and removed the rest of the wall. The only patching necessary here was to install 4½-in. wide strips of drywall along two walls and the ceiling, and to patch the flooring.

Removing ceiling sag — This trick involves the use of a strongback, but the principle would work for a header, too. I once worked on a motel that had two large dining rooms divided by a large hanging curtain. The curtain track was attached through the ceiling directly to the bot-

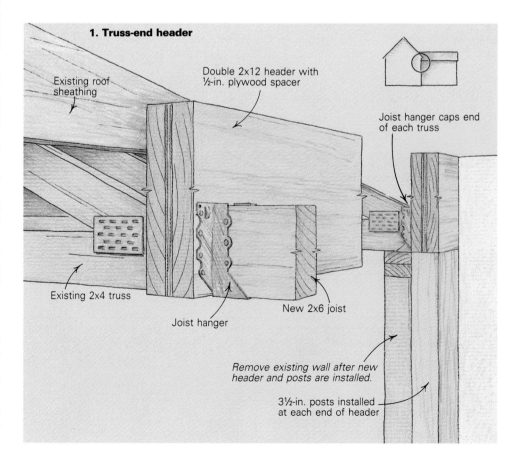

1. Truss-end header

Existing roof sheathing

Double 2x12 header with ½-in. plywood spacer

Joist hanger caps end of each truss

Existing 2x4 truss

Joist hanger

New 2x6 joist

Remove existing wall after new header and posts are installed.

3½-in. posts installed at each end of header

Drawings: Christopher Clapp

tom chord of one truss. Needless to say, the curtain bottom became wrinkled as gravity did its job and the noble truss sagged. The span was too long to double the truss up without adding extra support, so we went up in the attic and set a strongback beam perpendicular to the truss at the center of the span. Our beam also sat on three trusses on each side of the beleaguered curtain-bearer. We were told we couldn't use posts in the dining rooms, so we fastened the trusses to the beam with twisted metal straps, then fastened small cables to the beam and to the top chord of each truss. Structurally, this joined the beam, the top chord and the bottom chord into a single unit. Each cable included a turnbuckle somewhere along its length. After tightening the turnbuckles, we installed additional 2x framing between the beam and several top chords to lock everything in place.

Minimizing drywall damage—The final header method I'll discuss does result in a visible beam, but it saves the wallpaper. We needed to cut an opening in an existing wall. On the side of the wall that had painted drywall, we cut

that drywall where the top of the new header would be (drawing 4). On the wallpapered side, we cut the drywall out where the bottom of the new header would be. All the waste drywall was then removed, exposing the studs. We cut the nails holding the studs to the top and bottom plates, then carefully twisted each stud so that the drywall nails pulled out through the back of the wallpapered drywall. After setting posts in place at each end of the opening, we applied glue to one side of a 2x and slid this in place against the back of the wallpapered drywall. This invisibly fastened the wallpapered drywall to the header. We then set the second half of the header in place and *screwed* the two sides together to keep from disturbing the existing drywall. Wrapping the opening with wood trim left the wallpapered side as finished as ever.

There are other methods of hidden-beam installation, but I won't know what they are until the next owner gives us the next set of requirements and the wall is opened up.   □

Roger Gwinnup is a builder who recycles old houses near Iowa City, Iowa.

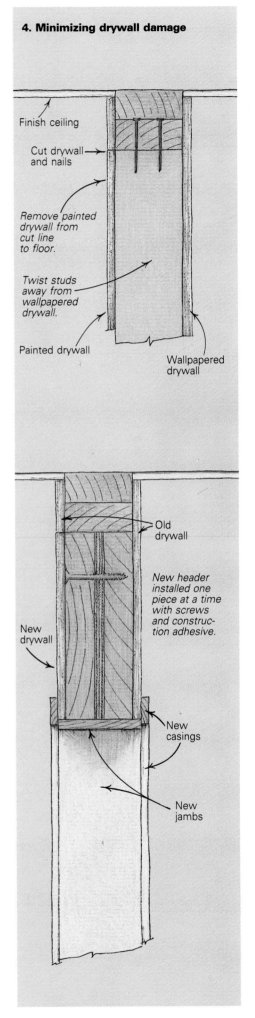

4. Minimizing drywall damage

Finish ceiling

Cut drywall and nails

Remove painted drywall from cut line to floor.

Twist studs away from wallpapered drywall.

Painted drywall

Wallpapered drywall

Old drywall

New header installed one piece at a time with screws and construction adhesive.

New drywall

New casings

New jambs

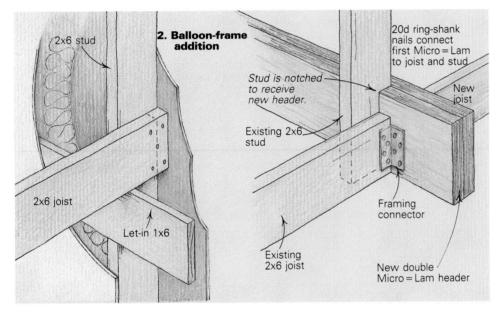

2. Balloon-frame addition

2x6 stud

2x6 joist

Let-in 1x6

20d ring-shank nails connect first Micro=Lam to joist and stud

Stud is notched to receive new header.

Existing 2x6 stud

New joist

Framing connector

Existing 2x6 joist

New double Micro=Lam header

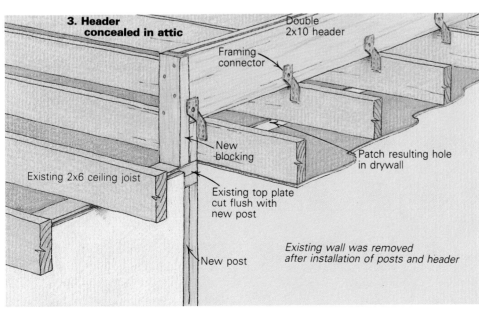

3. Header concealed in attic

Double 2x10 header

Framing connector

New blocking

Existing 2x6 ceiling joist

Existing top plate cut flush with new post

Patch resulting hole in drywall

New post

Existing wall was removed after installation of posts and header

Problem solvers

Concealed header

I needed to remove an 8-ft. bearing wall, but didn't want the new header to break the ceiling surface between rooms (drawing 5). Once the ceiling was shored up, the wall removed, and the joists exposed, I cut slots the thickness of doubled 2x10s, plus a spacer, through each joist. Because I was working alone, I pushed each 2x10 separately up through the gap, into the attic and supported each end on the top plates of opposing walls. Next, I connected each joist to the new header with joist hangers. Then it was a simple matter to patch the surfaces where the wall had been. Like magic, there was no trace of any structural element.

—*Steve Orton, a builder who lives in Seattle, Wash.*

Heavy long-span headers

Installing a long-span header is an inherently dangerous process. A 12-ft. to 15-ft. flitch-plate header or steel I-beam can weigh several hundred pounds. Here's how I easily and safely install a steel I-beam, using only one helper, some scrap lumber, and a couple of tools.

First I install two pairs of guides made of two 2x's between the ceiling and the floor at each end of the beam (drawing 6). The space between the guides should be about ¼ in. wider than the width of the beam. I then drill a ¾-in. hole through both sets of guides about 3 in. below where the lower edge of the installed beam will be. If the beam is heavy or I haven't had a good breakfast, I'll add some extra holes every foot or so. I use four sill-plate L-bolts or lengths of rebar in the holes for temporary beam supports. These allow me to lift one end of the beam and then the other as I "ratchet" it toward the ceiling. The advantage to this method is that the beam can't tip and I can stop any place I need to. To get the last two or three inches, I use a 4-ton hydraulic bottle jack and a scrap 4x4 post. This allows me to preload the beam enough so the cripple studs don't need to be bashed into place. Sometimes the building groans a little when I remove the supports from underneath the beam. The steel beam is usually straighter than the wall it replaces and the building has to adjust itself to the change.

—*Roy K. Jenson, a house remodeler in Edina, Minnesota.*

Lateral support for long-span headers

As a building inspector, I'm concerned about lateral movement in long-span headers. Fastening a header in place by toenailing it into the upper plate works for a short span, but I have my doubts about this technique for headers longer than 6 ft. to 8 ft., a span that is too great without some consideration for "racking." While this should be a design concern in every structure (to counteract wind loads, for example), it is of special concern in seismically active areas.

To solve this problem, I usually ask the builder to install ½-in. plywood on both sides of the wall at either end of the header (drawing 7). This is called a shear diaphragm. The plywood should extend uncut 12 in. to 16 in. onto the header from each end to form an inverted "L." Shear walls constructed of plywood must be a minimum of 5/16 in. thick for studs 16 in. o. c. Six-penny common nails, 6 in. o. c., are the smallest permissible size to be used in a shear panel. Nails at panel edges should be no less than ⅜ in. from the edge and no greater than 4 inches apart. Where a shear wall will not work, metal framing straps can be used in innovative ways to help provide lateral stiffening.

—*Lee Braun, a building inspector in Belvedere, California.*

Headers and point loads

One important effect to consider when installing long-span headers is that of newly created point loads. The load which was once evenly distributed along the bearing wall is now concentrated at two points. In many cases the original foundation may not be capable of supporting these concentrated loads without the risk of cracking or differential settlement. The existing footing and soil conditions should be investigated by an architect or engineer to determine the foundation's ability to support these new point loads. If there's a problem, the situation should be remedied *before* the point loads are created. I suggest that the structural situation be assessed by an architect or engineer when dealing with any span of 7 ft. or longer.

—*Martin Hammer, an architect in Oakland, California.*

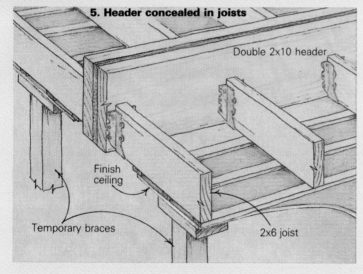

5. Header concealed in joists

Double 2x10 header

Finish ceiling

Temporary braces

2x6 joist

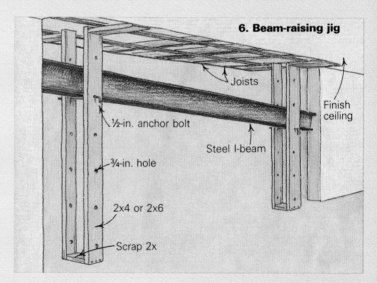

6. Beam-raising jig

Joists

Finish ceiling

½-in. anchor bolt

¾-in. hole

Steel I-beam

2x4 or 2x6

Scrap 2x

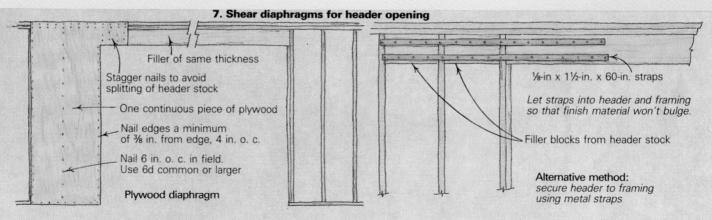

7. Shear diaphragms for header opening

Filler of same thickness

Stagger nails to avoid splitting of header stock

One continuous piece of plywood

Nail edges a minimum of ⅜ in. from edge, 4 in. o. c.

Nail 6 in. o. c. in field. Use 6d common or larger

Plywood diaphragm

⅛-in x 1½-in. x 60-in. straps

Let straps into header and framing so that finish material won't bulge.

Filler blocks from header stock

Alternative method: *secure header to framing using metal straps*

Stud-Wall Framing

With most of the thinking already done, nailing together a tight frame requires equal parts of accuracy and speed

by Paul Spring

I spent much of my former career as a carpenter building a reputation for demanding finishwork, but some of my best memories center around the sweaty satisfaction of slugging 16d sinkers into 2x4 plates as fast as I could feed the nails to my hammer.

The emphasis in framing is on speed. A lot has to happen in a short time. Accuracy, however, is no less important. The problems created by sloppy framing—studs that bow in and out, walls that won't plumb up and rooms that are out of square—have to be dealt with each time a new layer of material is added.

The fastest framing is done using a production system. But these techniques have long been the domain of the tract carpenter, and bring to mind legendary speed coupled with a legendary disregard for quality. However, production methods don't have to dictate a certain level of care. Instead, they teach how to break down a process into its basic components and how to economize on motion.

Done well, production framing is a collection of planned movements that concentrates on rhythmic physical output. It requires little problem-solving since most of the head-scratching has been done at the layout stage. As long as the layout has been done with care, a good framer can nail together and raise the walls of a small home in a few days, and still produce a house in which it's a pleasure to hang doors and scribe-fit cabinets. And this pace will give both the novice and the professional builder more time on the finish end of things to add the finely crafted touches that are rare in these days of rising costs.

If you know what the basic components of a frame are, nailing the walls together is simple. If not, you'll need to read the article on layout on pp. 25-33. After figuring out which walls get built first, you will separate the bottom and top plates (which were temporarily nailed together so that identical layout marks could be made on them); fill in between with studs, corners, channels, headers, sills, jacks and trimmers; and then nail them all together while everything is still flat on the deck or slab. The next step is to add the double top plate and let-in bracing. Finally you'll be able to raise the wall and either brace it temporarily or nail it to neighboring wall sections at corners and channels. Before joists or rafters are added, everything has to be *plumbed* and *lined*—this means racking and straightening the walls so that they are plumb and their top

Walls can be framed with surprising speed if they're laid out well. The pace isn't frantic— it's rhythmic, and based on coordination, economy of motion and anticipation.

plates exactly mimic the layout that was snapped on the deck—as discussed in the article on pp. 58-61.

For the sake of simplicity, I have stuck largely to giving directions for nailing together a single exterior 2x4 wall with most of the usual components. I have tried to mention how this process would be different under different circumstances, and how each section of the wall is part of a larger whole. If you are using 2x6s or have adopted less costly framing techniques, such as the ones suggested by NAHB's OVE (Optimum Value Engineering), you'll have to extrapolate at times from the more traditional methods explained here.

Getting things ready—The carpenter I apprenticed to would begin wall framing on a Thursday so he could recover over the weekend from those first two grueling days of keeping up with the hot young framers. You don't need to plan things to this degree, but

you do need to make sure that the right tools and materials are at hand. Leave the detailed plates tacked in place on the deck for the moment (or up on foundation bolts in the case of a slab). Make sure the deck is clean; if it's not, sweep it. This surface is going to be the center of your universe for the next few days, so keep it spotless and plan how you will use each inch of it.

While you are setting up on the deck or slab, a helper can be cutting headers to length if this hasn't already been done. If the ground is flat, set up on sawhorses; if not, use one corner of the deck or slab. A cutting list can be made directly from the layout on the plates. I usually number each door or window opening sequentially around the deck with keel (lumber crayon) on the top plates, and then use these numbers to identify the headers as I cut them. This way I can easily find the piece I need for a particular wall and snake it out of the pile of corners, trimmers, channels and headers stacked on the deck. The information for cutting rough sills, sill jacks, and blocking is right there on the layout too, and your helper can make up a package for each opening.

If you're working by yourself or with only one other carpenter, it's just as easy to cut this 2x material in place when you're framing. The only exception is trimmers, which can be counted up and gang-cut if the headers are all to be at the same height. If you aren't using standard 92¼-in. precut studs, now is also the time to cut studs to length. Gang-cutting them is easiest, but precision at this stage is still important. If a few studs cut short of the line happen to get nailed next to each other in a wall, a dip will be left in the floor above.

You should also count up the total number of corners and channels you will need, and nail these together up on the deck or slab. Many framers don't bother with this step. They nail their corners and channels together when they're framing, but pre-assembly avoids having to sneak edge nails into a channel that faces down and is crowded by a regular layout stud.

Before you litter the deck or slab with any more material, you first must figure out how much of the frame you are going to nail together before you raise some walls, and where you are going to begin framing. On some second-floor or steep-site jobs where the plan is very cluttered, it pays to stack the frame. This

means nailing together all of the walls and then raising them at one time. But this requires a lot of planning since the walls literally have to be built on top of each other (often three walls deep), and a big enough crew to lift and carry the walls into place. Big production jobs use framing tables—essentially huge wall jigs—or the flat ground around the house to complete all of the walls on one level before raising them. But usually it's best just to frame as many walls as your deck or slab will accommodate (figure that you can frame an 8-ft. wall if you've got at least 100 in. of room on the deck for its height), and then raise them, repeating this procedure until there aren't any sets of plates left.

The layout will have a lot to say about which walls get framed first. Exterior walls get priority, so you can save the precious working room near the center of the deck as long as possible. Of the exterior walls, you will be framing the by-walls first and then the butt-walls so that you can build as many walls in place as possible. Pick one of the longest exterior by-walls to begin; the back wall of the house is the traditional place to start.

Studs and plates—The best place for a lumber drop is right next to the slab or deck. This way you can literally grab a stud when you need one. But on a steep site or second story you'll need to pack the lumber to the deck as

you need it. A laborer will help in this situation, but don't give in to the temptation to stockpile studs on the deck—you'll only end up having to move them again. You can keep plate stock handy, though, by leaning it up against first-story framing. Spot two or three bunches of 10 to 20-footers around the building for double top plates. It's pretty easy to take the bows and crooks out of a double top plate when your'e nailing it, but if it's real bad, cull it out and start a pile that you can cut up into blocking and short jacks.

Nails—The only other items on the deck should be a skillsaw and a 50-lb. box of nails. The size and kind of nails a framer chooses seem largely regional. The allowable minimal size depends on whether you are *face-nailing* (through the face of one board into the face of another, such as nailing down the double top plate), *end-nailing* (through a board face into the end of board, such as nailing through the bottom plate into the studs) or *edge-nailing* (through the face of a board into the edge of another, such as a channel)—see the drawing, facing page, top. But rather than carry 8d, 10d, 12d, and 16d nails, it's easier just to carry a handful of eights in the small pocket of your nail bags for 1x let-in bracing and toe-nailing, and sixteens in the big pocket for everything else.

There are lots of choices when it comes to

nail coatings. Brights (regular steel nails without any coating) and even galvanized nails are okay, but I like *sinkers*. Sixteen-penny sinkers are a hybrid nail made in heaven (actually in Asia) for framers. Because of their coating, it takes half as many swings to drive one as an uncoated nail, and they don't crumple when you take a healthy swat. Their shank diameter is larger than a box nail, but not as thick as a common, which will split the ends of dry plate stock. They are also slightly shorter than 3½ in. (the length of the usual 16d nail), and have a thicker head. Sinkers can be either cement coated or vinyl coated.

In much of the far West, green vinyl (sold as g.v.) sinkers have become the predominant framing nail in the last five years. Although the vinyl does reduce the friction when they are being driven, these nails don't seem to offer much resistance on their way back out compared with other varieties. They're not nearly as bad, however, as nails that have had a treatment tract piece-workers affectionately call "gas 'n' wax." This is a coating made from kerosene and beeswax that is applied on site away from the eyes of building inspectors. You can whisper to these nails and they will drive themselves, but they unfortunately withdraw with the same ease.

My favorite nails are cement-coated (sold as c.c.) sinkers. The resin coating heats up from the friction of entering the wood and

Stocking the wall. Some framers stock the wall with studs before they split the plates apart (facing page), so they know how far back to carry the top plate. But separating the top plate and stocking it with headers first will define the openings right away and save having to flop heavy headers down in a sea of 2x4s. Either way, crowning the studs pays off.

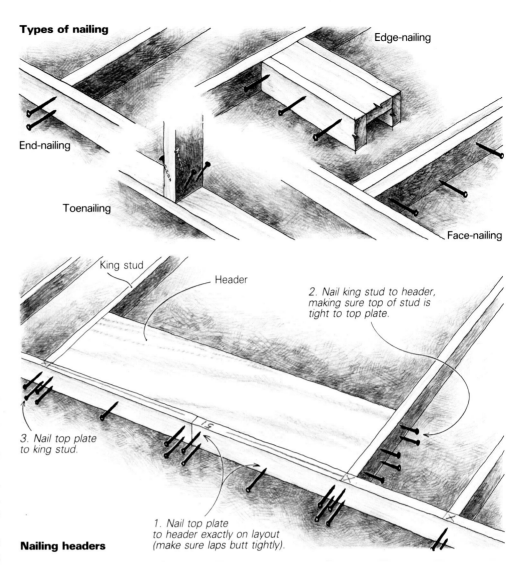

Types of nailing

Edge-nailing

End-nailing

Toenailing

Face-nailing

King stud

Header

2. Nail king stud to header, making sure top of stud is tight to top plate.

3. Nail top plate to king stud.

1. Nail top plate to header exactly on layout (make sure laps butt tightly).

Nailing headers

makes the nail slippery. But unlike the green vinyl sinkers, once the nail is in place, the coating bonds with the surrounding wood, and holding power increases several times over brights. The disadvantage of c.c. sinkers is that the black resin accumulates on your fingers, nailbags and hammerhead. If you're nailing finish boards with these sinkers (such as 2x tongue-and-groove roof decking over exposed beams), coat your hands with talcum powder before starting to keep from leaving black fingerprints all over the ceiling.

Building the walls—Unlike Jud Peake, who details plates flat in his system (see his article on pp. 25-33), I'm used to detailing plates up on edge. To frame a wall in place, I take the tacked-together plates and lay them just inside the snapped layout (wall) line with their inside edges down. Then I drive a 16d toenail through the bottom face of the bottom plate into the deck every 10 ft. or so. This way, when the top plate is separated, the bottom plate is already in position for framing. And once framed, the wall can be raised in place as if it were hinged to the floor.

Using this system, I separate the top plate with my hammer claw, walk it back on the deck and stock it with its headers first thing. This sets the location of wall openings early so you don't stud over one by mistake. It also means you don't have to flop a long, heavy header down in a sea of studs. But lots of other framers like to stock the wall with studs before separating the top plate so they know how far back to lay it (photo facing page).

Whichever order you use to stock the studs, cull the ones that are really crooked, and make sure that you lay all of the keepers crown up. Do this by sighting along each one before laying it down. Once you've got a stud on every layout mark, you can add corners and channels if you've made them up as units. Make doubly sure that the flat stud in each channel is facing as it should by looking at the layout on the floor as well as at the marks on the plate.

So far you haven't driven a nail, but soon that's all you'll be doing. If I'm using 4x12 headers and don't have to deal with head jacks, I like to nail the top plate to the top edge of the header first thing. This adds a lot of weight to the top plate and keeps it from moving around. Also. top-plate splices often come in the middle of a header, and you can begin making the wall a single unit by connecting the top plate to the header. Make sure the plates butt tightly so you don't lengthen the wall, and drive two nails into each plate end. At each end of the header, you'll also need two sixteens. In between, you should

stagger the nailing to each side at 16-in. centers. If the plate runs through without a splice, use two 16d nails at each end of the header, and then stagger nails every 16 in. in between, as shown in the drawing above).

Next, you should take care of the king stud to make sure that you have room to swing your hammer. Drive at least four nails through the face of each king stud into the end of the header. With a 4x12 header I use six 16ds. This is an important intersection. If the king stud doesn't stay tight to the header, it will pop sheetrock nails and leave a crack radiating away from the corner of door and window openings. Last, end-nail the top plate to the king stud with two nails. If you are using a header that doesn't reach the top plate and therefore requires head jacks, you can drive two nails into the top of each one through the top plate as you would a regular stud, and then toenail the bottoms to the top of the header with four 8d toenails each.

Now you can start nailing off everything that you've laid between the plates. Stay with the top plate and begin nailing at one end or the other. If you are right-handed, you'll find that working from left to right will probably be most comfortable and help you establish a rhythm. You'll be working bent over from the waist—one foot up on the edge of the plate, and the other foot nudging the stud onto the layout line and bracing it from twisting (photo

next page). Each 2x4, whether it's a layout stud or part of a corner or channel, gets two 16d nails driven into it through the plate.

Your first nail should be near the top edge of the plate, where the pencil layout line is marked. Set the nail with a tap of your hammer, then line the stud up on the mark and drive the nail through the plate and into the stud with your next couple of blows. Be careful to split the difference if the stud is narrower or wider than the plate to which you are nailing it. This may mean having to hold the stud up with your nail hand until you get the first nail in. Pay close attention to the layout; it's surprisingly easy to lose your concentration and begin nailing on the wrong side of the line despite the X on the edge of the plate that indicates the stud location. The only trick to the second nail is to make sure the stud is square to the plate. Judge this by eye.

Production framing requires a strong hammer arm and a dexterous nail hand. The only way to develop your arm is to drive a lot of nails, but there are some tricks to fingering the nails. With 16d nails, you need to orient the heads all in the same direction. I like to do this when I'm over at the nail box to refill my bags. This way, each time that you reach down you can pull out a large handful of nails ready to drive. Then, without dropping the nails cradled in your hand, use your index finger and thumb to reach into your palm to

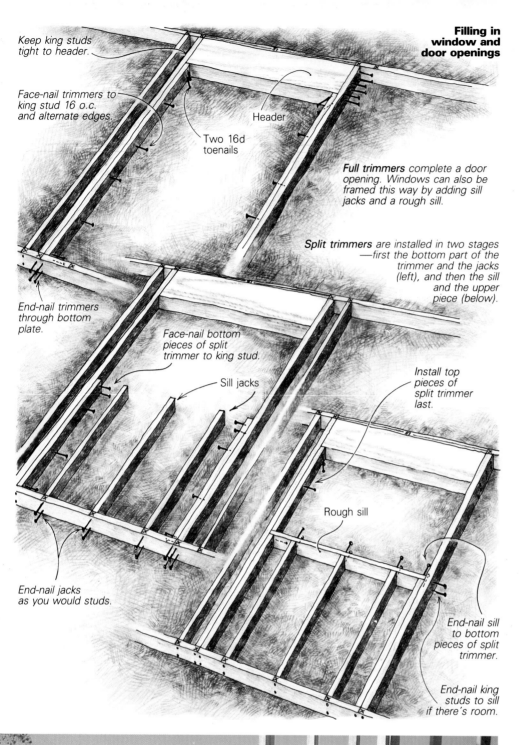

Keep king studs tight to header.

Face-nail trimmers to king stud 16 o.c. and alternate edges.

Header

Two 16d toenails

Full trimmers complete a door opening. Windows can also be framed this way by adding sill jacks and a rough sill.

Split trimmers are installed in two stages —first the bottom part of the trimmer and the jacks (left), and then the sill and the upper piece (below).

End-nail trimmers through bottom plate.

Face-nail bottom pieces of split trimmer to king stud.

Sill jacks

Install top pieces of split trimmer last.

Rough sill

End-nail jacks as you would studs.

End-nail sill to bottom pieces of split trimmer.

End-nail king studs to sill if there's room.

pinch a single nail. Extend the nail and rest it point down on the plate for the hammer while your other fingers regrip the nails in your palm. This should all happen while you are backswinging your hammer.

The first swing should be just a tap to start the nail. All carpenters hit their fingers occasionally, but you learn to keep your fingers out of the way when you are swinging hard enough to do any damage. This is particularly important with framing hammers, which can tear as well as bruise.

Once you've finished nailing off the top plate, move to the bottom plate and do the same thing all over again. You may want to reverse this process if you're framing exterior walls on a slab. The big problem here is that the anchor bolts invariably fall on the stud layout, and this requires chopping out some of the stud bottom. This is a lot easier if the top of the stud isn't already nailed.

When you are nailing the end studs on walls that butt by-walls on both ends, hold this last stud back from the end of both plates about ¼ in. This way if the stud bows out slightly, it won't prevent the wall from being raised. You can drive it back to its proper position when you nail the intersection together.

Another potential trouble spot is plate splices. If they come at a stud, you will need four 16d nails in the end of that stud—two from each plate end (photo below left). The other place that plates often join is in a doorway. Although the bottom plates will eventually be cut out, on long, heavy walls make sure that they stay butted by nailing a block on top of the plates at the joint.

Windows and doors—Once all of the headers, studs, corners, and channels are nailed in, you can complete the openings. Doors are easy. Fill in under each header with one trimmer on each side (two on each side if the opening is 8 ft. wide or more), as shown in the drawing at left. Trimmers need to fit snugly between the header and the bottom plate. You shouldn't have to pound the plate apart to get them in, nor should they be short. At its bottom, nail the trimmer like a stud. Then face-nail it to the king stud at 16 in. o. c., with the nails alternating from edge to edge. At its top, drive two 16s up at an angle to catch both the king stud and the corner of the header.

Taking some care with the trimmers will really pay off when it comes time to hang and case the doors. Make sure that the bottom of the trimmer is nailed right on its line—this will ensure a plumb and square opening. Also make sure that the edges of the trimmer and the king stud are even—this means that the

Framing. The top plate being nailed at left is a splice over a stud, which requires four nails driven on a slight angle. When space gets tight on the deck, two framers often end up working on the same wall. One should take the entire top plate; the other, the bottom plate. If they are both right-handed, they will begin on opposite ends of the wall since they will be moving left to right for comfort and speed.

mitered door casing will sit on a flat surface instead of a hump or dip. When you're finished nailing the trimmers, go back to the top of the king stud with your hammer and give it a bash or two so that it isn't separated at all from the end of the header.

Windows are a bit more complicated than doors because you've got the rough sill to contend with. Here you've got a choice. The sill can be cut to butt against both king studs, with *split trimmers* nailed up above and below it, or you can nail up full trimmers and toenail the sill into them. I was taught to use the split-trimmer method, and I prefer it because you can end-nail the sill into the lower trimmers and even end-nail through the king stud for a very solid connection.

If you're using split trimmers, begin by cutting the rough sill to length. Instead of measuring between the king studs, hold your sill stock up to the bottom of the header and mark it. Here the framing will be tight, so you'll get an accurate measurement even if the studs bow out down where the sill will be installed. Next, cut the sill jacks if their length is given in the layout, nail them at the bottom plate and drop the rough sill on top. If the sill-jack measurement isn't given, the size of the rough opening will be, and you can use your tape measure to mark the top of the rough sill. Allow 1½ in. for the sill, strike a line, and use this lower mark to cut in the sill jacks.

Remember that the lower part of a split trimmer is just another sill jack except that it also gets face-nailed to the king stud using 16d nails on alternating edges at 16 in. o. c. (drawing, facing page). Once all the jacks and lower trimmers are installed, you can end-nail the rough sill to them with two 16d nails at each intersection. If there's room, drive a couple of end nails through the outside of the king studs into the sill. If you can't do this, drive a 16d nail at a slight angle down through each end of the sill to catch the king stud.

The last step is to cut the top half of the split trimmer and face-nail it in. The only other complications are double sills and trimmers for very wide windows. In this case, install the first trimmer on each side full, and split the inside one around the doubled sill.

If you're not using split trimmers, you can treat the opening as you would a door at first by installing the trimmers full length underneath the header. Then mark the sill for length, and cut it. Follow the same procedure that was outlined above to set the sill and its jacks. But when you set the sill, use at least four toenails from the sill into the trimmer, along with a 16d toenail through the thickness of the trimmer into the end grain of the sill.

Rake walls and specials—Rake-wall (gable-end wall) framing is a little different because the tops of the studs have to be cut at the roof-pitch angle. But if the wall has been laid out right, cutting shouldn't slow you down too much. Mark the studs in place and cut their tops with the shoe of the saw set at the roof-pitch angle.

The bottom plate nails to the stud normally.

Blocking. Horizontal blocking for tubs and showers on interior walls goes faster if you let it in rather than cutting and nailing short blocks in the stud bays. Make sure to set your saw depth accurately to ensure that the blocking ends up in the same plane as the rest of the wall.

At the top plate, I usually drive a toenail through the toe (long point) of the angle at the top of the stud into the plate to hold the stud on the line. Then I drive two 16s down through the top plate into the stud, as you would in a standard wall.

The last step before double top plating any wall is to take care of the specials, which usually means different kinds of blocking. Interior soffits (drop ceilings) and walls over 10 ft. high will need fire blocking or stops—horizontal 2x4s nailed between studs to delay the spread of fire up the wall. Chalk a line across the studs and stagger the blocks on either side of the line with two 16d nails driven through the studs into each end of the block.

Mid-height flat blocking may also be required on exterior walls that will be covered with stucco. Production framers often raise this blocking a few inches on 8-ft. walls to 51 in. or 52 in. off the floor. This makes it easier to duck through the stud bays when working, more convenient as a shoulder support for lifting the wall onto foundation bolts on a slab, and keeps the blocks just slightly higher than electrical switch boxes.

Blocking for tubs and showers is a bit easier since it is usually installed with the face of the blocking nailed flush with the edges of the studs. It can either be cut in short blocks (14⅞₆ in. for 16-in. centers), or a length of 2x can be let into the studs (you can do this only on interior walls where the side that requires blocking is facing up). To let in blocking (photo above), position the 2x4 on the wall where the blocking will be needed and scribe

Bracing and raising. Let-in bracing can be cut into the top edges of the studs by setting the saw depth to 1½ in. and letting the shoe ride on the bracing (photo right). This procedure is not without risk of kickback. The safer alternative is scribing against the brace and cutting to the pencil mark. Facing page: The let-in brace in the foreground of the photo shows a forest of 8d nails that will be driven once the wall is raised and plumbed. Two keys to raising walls safely are even distribution of the weight and vocal coordination of effort.

Let-in bracing

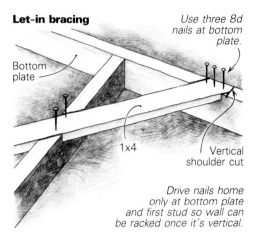

Use three 8d nails at bottom plate.

Bottom plate

1x4

Vertical shoulder cut

Drive nails home only at bottom plate and first stud so wall can be racked once it's vertical.

Double top plating

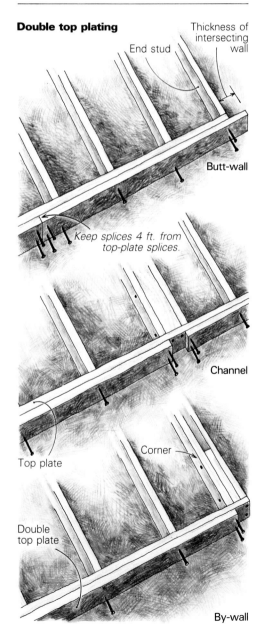

End stud

Thickness of intersecting wall

Butt-wall

Keep splices 4 ft. from top-plate splices.

Channel

Corner

Top plate

Double top plate

By-wall

a pencil line on either side. Cut to the inside of each of these lines with your saw set at 1½ in., remove the resulting scrap and nail the 2x4 in place.

Double top plate—The key to double top plating is the channels and corners. When two walls intersect, one of the double top plates acts as a tie between them. The double top plate on a butt wall will overhang its end stud 3½ in. for a 2x4 wall, and the double top plate of the corresponding by-wall will be held back 3½ in. to receive it. Double-plate splices should be held back at least 4 ft. from the end of a wall or from splices in the top plate.

Pull your plate stock up on the deck and lay it down where it will go. You may be tempted to use a tape measure to mark the overhang or hold-back on butt-walls. Instead, hold the stock to the far side of the channel and scribe the other end of the stock with your pencil held against the end of the wall. As long as the walls are all the same thickness, you will get a double top plate that's the right length by cutting at the pencil line.

Double top plating can go very fast. The ends of each piece require two 16d nails (drawing, left). In between you'll want to drive a 16d at every stud, alternating sides of the plate. Hitting the stud layout with these nails allows you to let in bracing in any of the bays without worrying about hitting a nail with your saw. You will only get to nail double top plates on butt-walls that are built out of place, since this plate has to project beyond its end stud by the thickness of the by-wall it will intersect. If it's easier to build the wall in

place, cut the plate anyway and nail it once the wall is raised.

Accuracy in cutting and nailing double top plates is essential. A double top plate that projects a little beyond a corner will drive you bananas when you're out on the scaffolding later nailing the sheathing. Be careful with the width of channels too—as soon as you leave a 3½-in. slot for the double top plate of an intersecting wall, it will invariably be cut from dripping wet stock and measure 3⅝ in. You can trim the double top plates on each side to make it work as it should, but you'll have to lay the wall back down to do it.

There's another precaution you should take just before raising a butt-wall that has been framed out of position. Give the overhanging double top plate a couple of blows from your hammer to drive it up off the top plate a half inch or so. This way when you are sliding the butt wall into place, the projecting double top plate won't hang up on the top plate of the already standing by-wall.

Bracing—All walls need some kind of corner bracing to prevent them from racking. There are lots of ways to get this triangulation. Sheathing and finished plywood siding provide excellent resistance if the nailing is sufficiently close. Cut-in bracing (flat blocks cut at an angle and nailed into each stud bay along a diagonal line) and metal X-bracing (long 16-ga. sheet-metal straps that are nailed to the studs under siding) are both quite effective. But for maximum strength when you are not using plywood, let-in bracing is usually specified. This 1x4 brace is mortised flat into

the exterior edges of the studs. It should extend from near the top corner of the wall down to the bottom plate at about 45°. In an 8-ft. wall with studs at 16 in., the brace will cut across six stud bays. You can get one into five bays by increasing the angle a bit.

Not every stud space needs to have a brace; the minimum standard for exterior walls and main cross stud partitions is one brace at each corner, and one at least every 25 lineal feet in between. But the more braces you use, the easier the wall will be to plumb and line, which in turn will create square rooms with plumb door jambs. And each wall that gets a let-in brace will act as a single unit, which in turn will increase the strength of your frame when it's subjected to wind or seismic loads. On a quality house, the small price you pay for 1x4 and the couple of minutes it takes to let it in allow you to brace even short walls. This goes for interior partitions as well. I even like to shear panel (⅜-in. plywood with close nailing) long cross walls.

A typical medium-length wall (say 30 ft. long) contains two let-in braces. These should form a V with an open bottom, since the tops of the braces start high on each corner. Of course it's not always this simple, because most walls contain window and door openings that have to be dodged.

These braces are best let in before the walls are raised, but you must make sure that the wall is close to being square. If you're framing the wall in place on the deck, it should be okay. But if it's been moved around some and is visibly racked, then you need to rough-square it by getting the diagonal measure-

ments approximately the same. Then lay a 1x4 across the top edges of the studs at an approximate 45° angle. Both ends should overlap the plates between studs. On an 8-ft. wall this will require a 12-footer. Now scribe each side of the 1x4 on every stud and at the top and bottom plates. Cut on the outside of each of these lines with your saw set at a fat ¾-in. depth. This will produce a slot about ¼ in. wider (two saw kerfs) than the brace—the extra width will accommodate the racking of the wall left and right during plumbing and lining. Remember, it's the number of nails and their shear strength that will keep the wall square eventually, not the fit of the brace in its slot.

Production framers eliminate scribing by cutting with the 1x4 in place (photo facing page). This procedure works best with a worm-drive saw, and requires a lot of experience with the saw because of the danger of kickback. Even then, safety takes a back seat to speed here. The brace is held down in place with one foot, and the saw is run along one side of it set to a depth of 1½ in. The shoe of the saw rides right on the brace. The framer will then change directions and saw back along the other side of the 1x4. The ends of the 1x4 are cut off in place at the bottom of the bottom plate and about ¼ in. shy of the top of the double top plate. The last step is to chop out the little blocks of wood between the saw kerfs.

Install the brace while the wall is still flat on the deck. Drop the brace into its slot, holding it flush with the bottom of the bottom plate, and drive three 8d nails there. You can also nail it to the first stud since this is still low

enough that it won't interfere with racking the top of the wall during plumbing and lining. You should also start two nails in the face of the 1x4 for each stud so that they can be easily driven home with one hand later. Start a total of five nails at the top of the brace—two into the double top plate and three just below this at the top plate. A lot of framers also start a nail in the top plate and bend it over the brace to keep it from flopping around when the wall is being racked.

Let-ins are most effective if they have a vertical shoulder at the bottom plate rather than coming to a point. This means making a plumb cut on the last inch or so of the brace measured along the angle, and a corresponding slot in the bottom plate (top drawing, facing page). Not all building inspectors will insist on this, but it's a good idea anyway.

If you're building on a slab, don't allow the let-in brace to sit on the concrete. End the brace in the middle of the 1½-in. thickness of the redwood or pressure-treated bottom plate.

Raising the walls—This is the best part, but also the point at which a lot of backs get wrecked (for the mechanical alternative, see sidebar, p. 94). Move anything you might trip over. If the wall is a rake, or is very tall or heavy, toenail the bottom plate to the wall line, or nail lumber strapping to the bottom of the plate, run it under the wall, and nail to the subfloor on the other side. For a standard 8-ft. wall, nail stops—short lengths of 2x4—to the joist just below the deck every 6 ft. to 8 ft. so that they stick up where the wall is going to be raised, preventing it from skidding over the

Mechanical wall raising

Nailing off the laps of the double top plates at corners and channels (facing page) ties the walls together. This doesn't prevent racking and bracing the walls plumb later.

Every carpenter has at least one story about a former partner or laborer who was 6 ft. 4 in. and immensely strong. However, being able to lift twice your body weight is of little consequence when you consider the weight of construction materials and the power of machines. While hiring a crane or forklift is usually not an economical option on small jobs, using simple mechanical advantage is, and it can change the way you work.

Wall jacks are a good example. With a pair of them, two people can easily raise a long wall full of solid headers weighing 2,000 lb. From there, the logical step is to sheathe exterior walls—and even add windows, siding, trim and paint—down on the deck before they are raised.

There are two kinds of wall jacks, but they work similarly. The first looks and operates much like a scaffolding pump jack. By pumping the handle of its ratchet winch (the same mechanism a car jack uses), the jack walks up a 2x4 (or 4x4), carrying the top of the wall to be raised with it on its horizontal bracket. In fact, these devices are often called walking jacks. The 2x4, which begins in a vertical position, is held from skidding by a block nailed to the deck. As the jack makes its way up, the 2x4 is allowed to get less and less vertical so that the wall will continue to bear on it. These jacks are relatively inexpensive, ranging from $60 to $125 apiece. Two brands I know of are Holtsma walking jacks (Box 595, River Street Station, Paterson, N. J. 07524) and Olympic Hi-jacks (Olympic Foundry, Box 80187, Seattle, Wash. 98108).

The other kind of wall jack (the one that I've owned) is manufactured by Proctor (Proctor Products Co., 210 8th St. South, P. O. Box F, Kirkland, Wash. 98033). It consists of a metal boom that is fitted to a hinged plate at the bottom that nails down to the subfloor and joists below. What amounts to a ¾-ton come-along is mounted on the boom just below waist height. The ⁵⁄₃₂-in. galvanized aircraft cable is threaded through a sheave (pulley) at the end of the boom and is fitted with a nailing bracket that attaches to the double top plate of the wall to be raised. The boom begins in a vertical position, and begins to lean as soon as the ratchet winch is put to use and the wall begins to come off the deck. An

adjustable stop prevents the wall from going beyond vertical. Proctor wall jacks come in three lengths (16 ft., 20 ft. and 23 ft.) and all of them telescope for carrying or storage. Although this system is more expensive— the smallest pair of jacks retails for $445— it's very safe and can handle walls as long as 75 ft. with just two carpenters on the job.

The real advantage to wall jacks is that they give you the freedom to finish a wall completely while it's flat on the ground. This means money saved even if you're building a one-story house on a flat lot. On high work, you often can eliminate the cost of scaffolding and speed the usually slow progress of sheathing, windows, siding, trim and paint by doing most of the work right where the frame sits on the deck. On the houses where I've chosen to do this, I've also been able to use less skilled labor to complete the walls. For instance, any careful person can apply a full-bodied stain with a roller when the wall is flat and accessible. It doesn't have to be sprayed by a painting sub. Trimming out windows can be left to an apprentice, since having to recut a piece for fit isn't a big deal when you're not climbing down off scaffolding to do it.

Tilting up finished walls isn't for every house. It is best used on modest rectangular plans with long walls where the siding and the trim detailing are simple. In any case, corners have to be finished off from ladders or simple wood scaffolding suspended from nearby window sills.

The only real trick to completing a wall on the deck is to rack the framing so it's exactly square before you sheathe it. This is worth checking several times. The alternative is spending half an hour with a cat's paw removing all that shear nailing so that the ends of the wall will sit plumb once in place. To check for square, toenail the bottom of the bottom plate of the wall to the deck right on the snapped wall line and then use a sledgehammer to bump the top of the wall until the diagonals measure the same. The only other complication is with butt-walls once the by-walls are up. Butt-walls have to be finished slightly out of position (left and right) to buy clearance for the sheathing or siding that will overlap the corner framing of the by-wall, or the end panel will have to be left off until the wall is raised. —P. S.

edge. Also nail a long 2x4 with one 16d nail high up on each end of a by-wall. The single nail will act as a pivot so that the bottom of the brace will swing down and can be angled back to about 45° alongside the deck. It can then be nailed to the rim joists when the wall has been raised to approximately plumb. More brace material should be stacked nearby so that it can be grabbed quickly.

To get the wall in a position where you can get your hands under it, lean short lengths of 2x against the face of the double top plate every 12 ft. or so. Then, standing inside the wall itself, bury your hammer claw in the double top plate with a healthy swing. When you lift the top of the wall just a few inches, the blocks will fall beneath the top plate and you'll have enough room to get a grip.

Now it's time to gather your crew. Most carpenters can lift a good 12 lineal feet of 2x4 framing—more if there aren't a lot of 4x12 headers, less if the wall is 2x6 or framed with very wet lumber. Spread people out along the wall according to where the weight is. Headers are the worst because the weight is all at the top. The ends of the wall are almost always the lightest. The first maneuver in lifting the wall (photo previous page) is called a clean and jerk in weightlifting. If you don't bend your legs it's a sure road to a hernia or a bad back. The second stage—where you've got the wall to your waist and you are pushing with the palms of your hands—is basically a press and should be done with your legs braced behind you. Don't make it a contest. Raising a heavy wall requires staying in sync with everyone else.

If you're raising a by-wall, at least one of the crew should let go once it's up—with fair warning—and nail the outside braces. The wall should lean out slightly at the top to leave a little extra room for the butt wall that will intersect it.

Raising walls on a slab is slightly different. Here you've first got to raise the wall to a vertical position, and then lift and thread the bottom plate onto the foundation bolts. Long 2x4s on edge can be used effectively as levers under the bottom plate to lift the wall up above the bolts, as long as you have someone steadying the top of the wall. End braces can be nailed off to stakes driven in the ground right next to the slab. Once the wall is steady, beat on the sill at several spots to make sure that it's down, then put washers and nuts on the foundation bolts and screw them down finger tight.

If you're raising an exterior butt wall on either a slab or a deck, you'll be nailing the end stud to the corner of its corresponding by-wall, rather than using a brace. Make sure that the bottom plate is flat on the deck and that the two top plates match in height. Also align the outside face of the corner and the edge of the end stud so that they are in the

same plane all the way up. Alternate 16d nails on each side of the end stud every 16 in. If you are raising a partition or interior wall, nail its end stud to the channel in the same way, with the same kind of care.

Long walls may require an intermediate brace or two before everyone can let go. How much bracing you need to add depends in part on how soon you'll be going home for the day. Braces take up a lot of space on the deck when you're trying to frame the rest of the walls. But when you leave the site, it's a whole different story. Figure that a hurricane will strike that night and brace everything off accordingly, especially if you have already sheathed your walls.

If you're bracing off walls on a concrete slab for the night or weekend, you can use a "tepee" on exterior walls. To make one, take a 14-ft. piece of plate stock and run it through a stud bay of the wall to be braced so that half

of the plate stock is cantilevered out beyond the building, and the other half is on the inside of the slab. Then nail a 2x4 brace from the top of the wall down to the end of the plate stock on both the inside and outside. This kind of brace allows a little give, but the wall won't go anywhere.

A precaution you want to take on a plywood deck is to nail down the bottom plate. First make sure that the ends of the wall are where you want them—on a layout mark, flush with the perimeter of the deck, or butting another wall. Then use a sledgehammer to persuade the bottom plate into a straight line that sits just at the edge of the wall line established in chalk during layout. After that, drive one 16d nail per bay to keep it there.

The last thing that you have to do to connect the walls is to nail down the double top plates that lap over intersecting walls. Remember that the walls don't have to be

plumb; you'll take care of that later. You are just making sure that the walls are nailed together exactly as they were laid out from the bottom plate all the way up.

To nail off the double top plates that lap, you can claw your way up a corner and walk the plates between channels. But you'll probably get it done more safely by moving a 6-ft. stepladder around. First, make sure that the walls are driven together tightly all the way up, and that they are aligned with each other vertically. Then you can finish off the double top-plate nailing using four 16d nails for each 2x4 plate lap.

When all the plate pairs are nailed down, your frame will be complete. Plumbing and lining it will make it ready for joists or roof rafters. The feeling at the end of a day of this kind of work is unparalleled. You are surrounded by the tangible evidence of your progress and the worth of your labor. □

Installing a Long-Span Header

How to open up an existing bearing wall

by Matt Holmstrom

One of the most dramatic ways to alter a living space is to tear out a wall or cut a large opening in it. "Open up the floor plan" and "let more light in" are catch phrases every remodeler hears frequently. But many homeowners and novice remodelers find this an intimidating prospect. The work is extremely messy and seems chaotic. And two words— "bearing wall"—keep many people from tackling it. Actually, this is a straightforward job that demands more common sense than technical skill. Whatever the circumstances, the same basic procedure is followed. Advance planning, careful observation and a step-by-step approach will eliminate that "I'm in over my head" panic.

This view is from the newly framed addition toward what had been the exterior wall of the house. My clients wanted the wall opened up in order to turn the rooms into a single space. Note the various explorations that offer clues as to what lies beneath the sheathing.

Know thine enemy—Simply put, a bearing wall is a wall that bears some of the weight of the structure above it. A wall that is not load bearing supports only its own weight and that of the finish materials on it. To remove a non-bearing wall, just demolish it. Removal of a bearing wall, however, will require some temporary structure (called shoring) to support loads while the work is going on, and will require the installation of a permanent load-bearing member (usually a header or beam supported by posts or studs) to take the place of the removed portion of wall. If you expect to have a framed opening by afternoon where there was a wall that morning, much of the work will have to be done before the reciprocating saw comes out of its case.

First, determine if the wall to be altered is in fact a bearing wall. This is usually fairly easy. Generally, a bearing wall is perpendicular to the joists and/or rafters above it. The weight supported by the joists is sitting on the wall. Another clue is to look beneath the wall (in the basement or the crawl space) for indications that the wall is transferring loads to the ground. You might find another framed wall, a beam set on piers or a foundation wall. If the suspect wall is on a second floor, you'll have to figure out what supports it and then look to see what's beneath *that* wall.

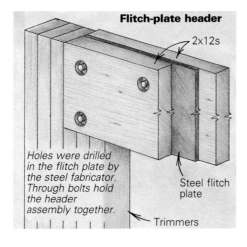

Flitch-plate header

2x12s

Holes were drilled in the flitch plate by the steel fabricator. Through bolts hold the header assembly together.

Steel flitch plate

Trimmers

Sizing the header—The next step, assuming that you are indeed dealing with a bearing wall, is to figure out approximately how much wall you'll need to remove and then size the header. The carpenter's rule of thumb used to be this: for spans 4 ft. or less, the header was made of doubled 2x4s; for up to 6 ft. of span, doubled 2x6s; and so on up to 12 ft. of span and 2x12s. Nowdays, however, the building inspectors in my area always want to see at least 2x10s in a bearing-wall header, so I use these for anything up to 10 ft. Given that the structural integrity of the house depends on

correctly sizing this header, I'd recommend that you check with your local building department if you're at all in doubt. In any case, headers less than 12 ft. in length don't call for anything fancy, and you can get the materials at the local lumberyard.

When you're dealing with header spans greater than 12 ft., or if the header will be supporting unusual loads (a large bathtub, perhaps, or a slate roof), you'll need to plan extra carefully. The options for these headers include a steel flitch plate bolted between faces of 2x stock, a steel I-beam, a glue-laminated beam, or perhaps a truss. Cost, availability, delivery time, weight and the available headroom are all factors to consider. A structural engineer or an architect can help you with this decision. In fact, your local building official may require that you consult an engineer or architect before proceeding. With any of these manufactured headers, exact span measurements are crucial—you don't want to trim ½-in. steel plate at the job site if you don't have to. Also, you'll have to allow additional lead time to obtain the header.

Checking the wall—The last part of your preliminary work is to check the wall for mechanical systems and to coordinate with the proper subs if necessary. Wiring, plumbing, and heat ducts may be in the wall and, with a few exceptions, will have to be eliminated or rerouted before the header can be installed. Once you start removing structural members, you can't dilly-dally around waiting for the electrician to show up. If you have the luxury of working in an unoccupied house, you can strip off the wall surfaces now and find out what you're dealing with. Otherwise, some detective work will be necessary.

Water pipes and heating ducts are pretty easy to track. If there is much plumbing in the wall, you've picked the wrong wall to tear out—it gets to be quite a job. As for wiring, wall outlets in the work zone are not always a problem. If wires come up through the sole plate, outlets can be left alone until the structural work is completed. But any wire that comes down through the top plate will have to go right away; that usually means switches will have to be relocated.

A case study—Once all the preliminary work is done, you're ready to proceed with wall removal. By now you know approximately what size opening you are going to cut in the bearing wall; the header materials are on hand; and you have dealt with, or are prepared to deal with, any mechanicals in the wall. Things go quickly now. In one or two working days, depending on the complexity of the

With the header in place, trimmers were quickly positioned and nailed to the king studs. Then the header was toenailed to the framing.

job, you'll have a new opening ready to finish. I recently opened up a bearing wall and replaced a good portion of it with a header. Here's how it worked out.

We were called in to build an addition to a brick-veneer ranch house. To make the month-long project easier on our clients, we removed a portion of the brick veneer, built the addition, then removed the load-bearing wall between old and new (photo facing page). That way, the exterior wall was never opened to the outdoors. Because the house had a hip roof, all exterior walls were bearing walls. In addition to supporting the old roof rafters and ceiling joists, the wall we removed would have to support the new ceiling joists of the addition.

In order to open the dining room to the new addition, we would have to replace most of the bearing wall with a 15-ft. header and appropriate support framing. One end of the header would rest on new studs added to what remained of the original wall; the other end would sit in a framed pocket formed by the junction of the old wall and the addition wall. One framing concession to our later tie-in had been building the floor of the addition slightly off level to match the existing floor; we didn't want to draw attention to the juncture of old and new.

For the header I opted for a ½-in. steel flitch plate bolted between a pair of 2x12s (drawing facing page). The plate cost about $150, and our local steel fabricator had it ready on just a few days' notice. Because the finished opening would be approximately 15 ft. wide, I ordered the plate 15-ft. 9 in. long. That would allow 4½ in. of support under each end, which is what our building inspector asked for. The 11-in. width of the plate would allow us to fit it completely within the depth of the 2x12s.

The wall contained a few switches, an outlet, an exterior light fixture and two hot-air supply ducts in the portion we planned to remove. The electrician eliminated all these wiring circuits when he roughed in the wiring for the addition. If you plan to leave any wiring in a wall during demolition, however, find the panel box and shut off all circuits to the area before beginning demolition. As for the hot-air ducts, we decided to disassemble them once the wall was stripped; the HVAC guy would later reroute them to supply the new addition.

Stripping the wall—It's incredible how much mess and debris even a small demolition job creates. I try to isolate the work area from the rest of the house and minimize the mess as best I can. Masking tape held 6-mil plastic sheeting over every opening that led to the rest of the house. If there are appliances or large pieces of furniture that can't be moved from the work area, I cover them with plastic sheets or drop cloths. I tape red rosin paper over nearby finished floor surfaces because plastic sheets are just too slippery—they're not tough enough, either. Besides, the paper is cheap and fairly tough, and the 3-ft. wide rolls are easy to handle. To absorb direct hits from dropped tools and falling debris, I lay a sheet of plywood over any finished floor adjoining the wall. Now's the time to remove all trim, doors, hardware or anything else that you want to save. Make sure you store them in another location, too. Once you start tearing into the wall, a certain inertia of demolition takes over and anything can just disappear in the debris.

This house had fiberboard sheathing on the exterior side of the wall to be removed and plaster over rock lath on the interior. I drew a rough layout of the opening directly on the plaster, allowing more than enough length for the header and a few studs ganged on each end; this determined where to cut the plaster. After removing trim and the existing window and door, we tackled the work. We stripped the plaster right up to the ceiling along the entire length of the header, so we had a good view of the doubled top plate our header would be supporting. After cutting plaster, we always clean up the mess to avoid grinding plaster dust into the finish flooring (oak in this case); protective paper can't always contend with the fine, gritty powder left by this kind of demolition (that goes for drywall demolition, too). A shop vac is almost a necessity here.

Building the header—With the wall framing exposed (but not yet cut), the new header, king studs and trimmer studs can all be laid out, and the header can be built. A header of this length (15-ft. 9 in.) and weight needs three trimmer studs on each end for support. Normally, if the exact position of the new wall opening is not critical, I try to use an existing stud as one king stud for the new header, and I begin my final layout off this.

Temporary 2x shoring in the foreground of the photo below was placed to support the ceiling loads before the structural portions of the bearing wall just behind it could be removed. Lifting the flitch-plate header into place called for plenty of manpower and some well-choreographed moves. The left end of the header fits into a framed pocket in an existing wall, while the right end will be supported on new 2x framing.

Here, the kitchen-window stud was my starting point. This was 6⅞ in. back from the drywall face of our perpendicular addition wall. I sistered a new stud against this old one, shimming between them to get it plumb. This would be the king stud for the new header. The remaining 5⅜ in. between this stud and the intersecting wall face could be filled nicely with three trimmer studs (4½ in.), a ⅜-in. filler of plywood or drywall and ½-in. drywall. The result would be a nice outside corner where the walls intersect. Then I toenailed the opposing king stud into place 189 in. away from the first king stud. I left the bottom plate in place for now.

The toughest part of this project turned out to be moving the steel flitch plate. It weighed 295 lb., and it took two men one-half hour to slide the plate off the truck racks and maneuver it onto sawhorses without damaging fingers or backs. Assembling the header was comparatively easy. I had hand-picked two straight 16-ft. 2x12s at the lumberyard. You want as little crown as possible in any header, but here even ⅛ in. of crown would have been difficult to deal with when it came to fitting the header into position. Any problems in fitting such a long, heavy header could mean more than wasted time—it could result in personal injury.

The flitch plate had been predrilled by the steel fabricator, so assembly was simply a matter of marking the hole alignment on the 2x stock, drilling and countersinking the holes, and securing the whole affair with ⅜-in. by 3½-in. bolts, nuts and washers. We lugged the completed header into the addition and set it at the base of the stripped wall. The only way two men could move this monster was by

sliding it along "leapfrogged" sawhorses up to the door, and then sliding it along the addition subfloor. Before we could begin to remove the last of the old wall, however, we had to shore up the ceiling to support ceiling and roof loads.

Setting the shoring—There are two ways I know of to build shoring, and I used both on this job. One method calls for building a 2x6 stud wall to support loads; we did this in the addition (top photo, previous page). Line up the studs under every second joist you have to support (every 32 in. o. c. in this case). The second shoring method calls for a beam (two sistered 2x6s) and two or three posts to support it against the ceiling loads; we did this in the dining room. The first method takes longer to build, but is probably more stable than the second. It spreads the load better, too. I use the second method under an uneven plaster ceiling—it minimizes ceiling damage, and I don't have to spend time locating ceiling joists in the plaster. To be effective, the shoring must be snug against the ceiling, but not so snug that it causes damage.

I usually use 2x6s for shoring walls. They're a little more rigid than 2x4s, and less likely to split when you bang them into place. The shoring should be set about 2 ft. from the bearing wall so that there's plenty of room to maneuver a stepladder between shoring and wall.

After the shoring was in, we completed the demolition by cutting away the rest of the wall studs. The easiest way is to cut each stud in half with a reciprocating saw and remove it, then cut the nails protruding from

the top and bottom plates. Use bi-metal blades in your reciprocating saw (I buy ones labeled: "For nail-embedded wood"); they cost more but last much longer for this kind of work. If you bend a blade (and you will), simply straighten it with pliers and get back to work.

Installing the header—You'll often be working in a small, cluttered area, so choreograph the installation: who will be where, which end of the header will go in first. I used a disc sander to bevel slightly one end of the header along its width. This made it considerably easier to slide the header between the king studs. After recruiting help in the form of two carpenter friends who were working nearby, I leaned the precut trimmer studs against the wall near each end of the opening. The rest happened fast: the four of us (with a bit of help from a friend) lifted the header up and slipped one end in first, while a man on each end knocked the first trimmers in place and quickly tacked them to the king stud (bottom photo, previous page). Then the header was driven the rest of the way in with a 3-lb. hammer and a scrap block, and an additional two trimmers were nailed off on each end, with two more studs behind the king stud. After toenailing the header into the top plate, we cut out the bottom plate flush with each end trimmer and pulled up for a rest. Our opening was framed and would be ready to finish once the shoring came down. ☐

Matt Holmstrom is a remodeling contractor who prefers to work on older homes. Photos by Bill Hoy.

The effect of removing a wall is dramatic. In this case, the original front door was directly in front of this basement door.

Framing Headers and Corners

A summary of framing details for strength and energy efficiency

by Rob Thallon

The walls of a building serve several important purposes: They define the spaces within a building to provide privacy and zoning, and they enclose the building itself, keeping the weather out and the heat or cold in. Walls provide the vertical structure that supports the upper floors and roof of the building, as well as the lateral structure that stiffens the building. Walls also enclose the mechanical systems (electrical wiring, plumbing and heating). To incorporate all of this within a 4-in. or 6-in. deep wood-framed panel is quite an achievement, and numerous decisions need to be made in the course of designing a wall system for a wood-frame building. There are two preliminary decisions to make that establish the framework for the remaining decisions. Once these are made, details such as headers and corner posts can be determined.

Wall thickness—Should the walls be framed with 2x4s or 2x6s? The 2x6 wall has become increasingly popular in recent years, primarily because it provides more space for insulation. This advantage comes at some cost, however. A 2x6 wall with studs spaced 24 in. o. c. (the maximum spacing allowed by codes) uses about 20% more material for studs and plates than a 2x4 wall with studs spaced 16 in. o. c. On the outside of the wall, the sheathing has to be ½ in. thick (⅛ in. thicker than sheathing on a standard 2x4 wall), and inside, the drywall also has to be ⅛ in. thicker to span the greater distance between 2x6 studs. Thicker insulation costs more, too. So, overall, 2x6 framing makes a superior wall but one that costs more. Framing the exterior walls with 2x6s and interior walls with 2x4s is a typical combination when the energy-efficient 2x6 wall is selected. Stud spacing of 2x4 and 2x6 walls may vary with loading, lumber grades and finish materials.

Framing style—Should the walls be built using platform framing or balloon framing? Balloon framing, with studs continuous from mudsill to top plate and continuous between floors, was developed in the 1840s and is the antecedent of the platform-framed wall. In recent years, balloon framing has been almost completely superseded by the more labor-efficient and fire-resistant platform-frame construction.

Headers and corners—Header size depends on wood species and grade, loading, header design and rough-opening span. Following is a rule of thumb for sizing a common header type, the 4x header:

For a single-story building with a 30-lb. live load on the roof and 2x4 bearing walls, the span in feet of the rough opening should equal the depth (nominal) in inches of a 4x header. For example, openings up to 4 ft. wide require a 4x4 header. The drawings on the following pages show various commonly used solutions to the problems of tying adjacent walls together and providing weight transfer around openings. □

Rob Thallon is an architect and builder in Eugene, Oregon. His book, Graphic Guide to Frame Construction, *from which this article is excerpted, was published by The Taunton Press in 1991.*

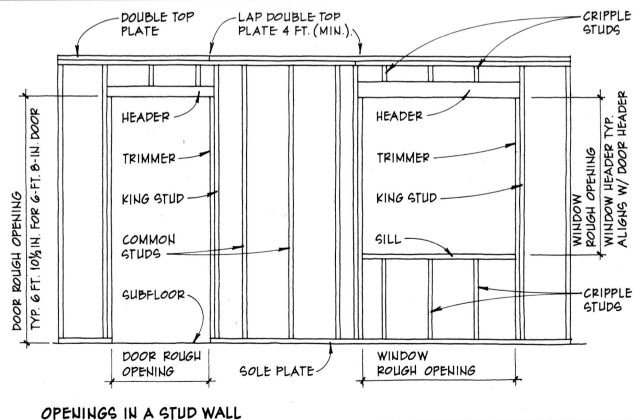

OPENINGS IN A STUD WALL

Framing Headers and Corners

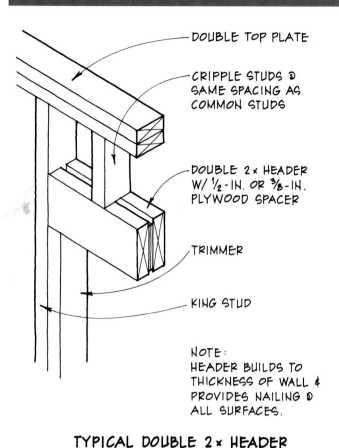

DOUBLE TOP PLATE

CRIPPLE STUDS @ SAME SPACING AS COMMON STUDS

DOUBLE 2× HEADER W/ ½-IN. OR ⅜-IN. PLYWOOD SPACER

TRIMMER

KING STUD

NOTE: HEADER BUILDS TO THICKNESS OF WALL & PROVIDES NAILING @ ALL SURFACES.

TYPICAL DOUBLE 2× HEADER
2×4 BEARING WALL

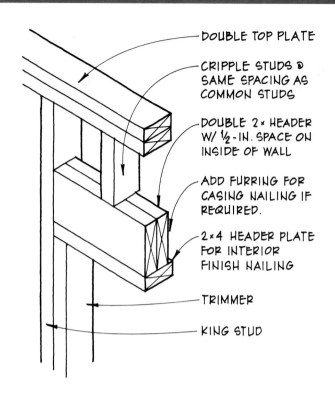

DOUBLE TOP PLATE

CRIPPLE STUDS @ SAME SPACING AS COMMON STUDS

DOUBLE 2× HEADER W/ ½-IN. SPACE ON INSIDE OF WALL

ADD FURRING FOR CASING NAILING IF REQUIRED.

2×4 HEADER PLATE FOR INTERIOR FINISH NAILING

TRIMMER

KING STUD

ALTERNATIVE DOUBLE 2× HEADER
2×4 BEARING WALL

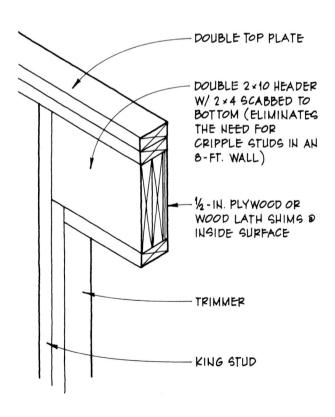

DOUBLE TOP PLATE

DOUBLE 2×10 HEADER W/ 2×4 SCABBED TO BOTTOM (ELIMINATES THE NEED FOR CRIPPLE STUDS IN AN 8-FT. WALL)

½-IN. PLYWOOD OR WOOD LATH SHIMS @ INSIDE SURFACE

TRIMMER

KING STUD

DOUBLE 2×10 HEADER
2×4 BEARING WALL

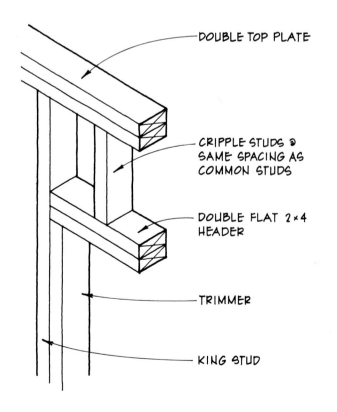

DOUBLE TOP PLATE

CRIPPLE STUDS @ SAME SPACING AS COMMON STUDS

DOUBLE FLAT 2×4 HEADER

TRIMMER

KING STUD

FLAT 2×4 HEADER
2×4 PARTITION WALL

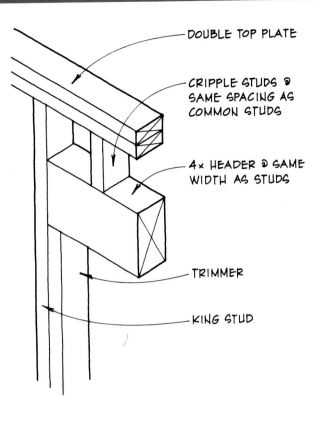

- DOUBLE TOP PLATE
- CRIPPLE STUDS @ SAME SPACING AS COMMON STUDS
- 4× HEADER @ SAME WIDTH AS STUDS
- TRIMMER
- KING STUD

4× HEADER
2×4 BEARING WALL

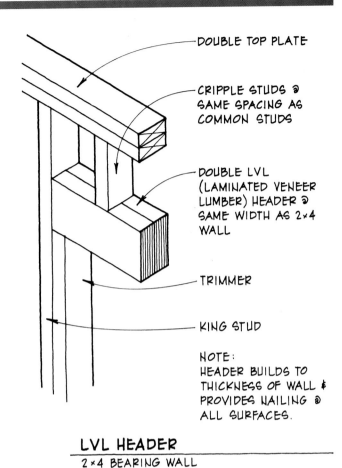

- DOUBLE TOP PLATE
- CRIPPLE STUDS @ SAME SPACING AS COMMON STUDS
- DOUBLE LVL (LAMINATED VENEER LUMBER) HEADER @ SAME WIDTH AS 2×4 WALL
- TRIMMER
- KING STUD

NOTE:
HEADER BUILDS TO THICKNESS OF WALL & PROVIDES NAILING @ ALL SURFACES.

LVL HEADER
2×4 BEARING WALL

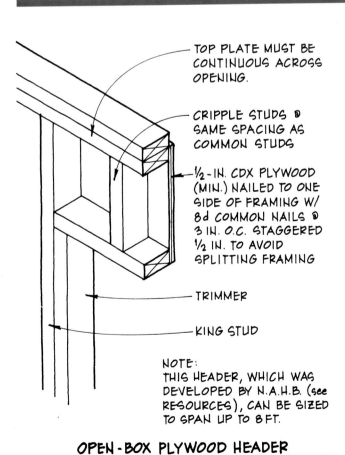

- TOP PLATE MUST BE CONTINUOUS ACROSS OPENING.
- CRIPPLE STUDS @ SAME SPACING AS COMMON STUDS
- ½-IN. CDX PLYWOOD (MIN.) NAILED TO ONE SIDE OF FRAMING W/ 8d COMMON NAILS @ 3 IN. O.C. STAGGERED ½ IN. TO AVOID SPLITTING FRAMING
- TRIMMER
- KING STUD

NOTE:
THIS HEADER, WHICH WAS DEVELOPED BY N.A.H.B. (see RESOURCES), CAN BE SIZED TO SPAN UP TO 8 FT.

OPEN-BOX PLYWOOD HEADER
2×4 BEARING WALL

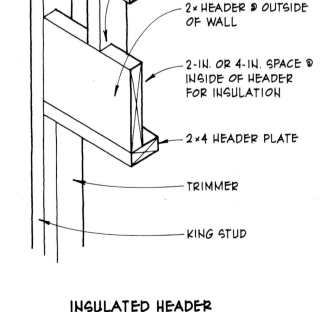

- DOUBLE TOP PLATE
- NOTCH CRIPPLE STUDS FOR 2× HEADER.
- 2× HEADER @ OUTSIDE OF WALL
- 2-IN. OR 4-IN. SPACE @ INSIDE OF HEADER FOR INSULATION
- 2×4 HEADER PLATE
- TRIMMER
- KING STUD

INSULATED HEADER
2×4 OR 2×6 EXTERIOR WALL

Framing Headers and Corners

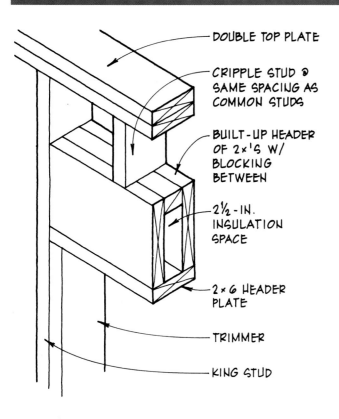

DOUBLE TOP PLATE

CRIPPLE STUD @ SAME SPACING AS COMMON STUDS

BUILT-UP HEADER OF 2×'S W/ BLOCKING BETWEEN

2½-IN. INSULATION SPACE

2×6 HEADER PLATE

TRIMMER

KING STUD

INSULATED DOUBLE 2× HEADER
2×6 BEARING WALL

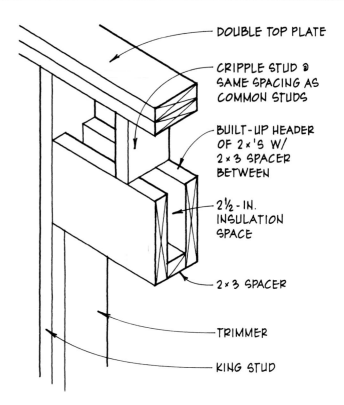

DOUBLE TOP PLATE

CRIPPLE STUD @ SAME SPACING AS COMMON STUDS

BUILT-UP HEADER OF 2×'S W/ 2×3 SPACER BETWEEN

2½-IN. INSULATION SPACE

2×3 SPACER

TRIMMER

KING STUD

INSULATED DOUBLE 2× HEADER
2×6 BEARING WALL/ALTERNATIVE DETAIL

CORNER STUDS BUILT UP W/ 2×4 BLOCKING BETWEEN PROVIDES NAILING @ INSIDE CORNER.

SOLE PLATE

NOTE: THIS DETAIL WORKS FOR BOTH INSIDE & OUTSIDE CORNERS.

2×4 CORNER
W/ BLOCKING

EXTRA STUD ADDED PERPENDICULAR TO CORNER STUD PROVIDES NAILING @ INSIDE CORNER & ALLOWS SPACE FOR INSULATION @ CORNER.

2×4 STUDS @ 16 IN. O.C.

SOLE PLATE

NOTE: THIS DETAIL WORKS FOR BOTH INSIDE & OUTSIDE CORNERS.

2×4 CORNER
W/ INSULATION @ CORNER

DOUBLE TOP PLATE OVERLAPS
@ CORNERS TO LOCK TWO
WALLS TOGETHER.

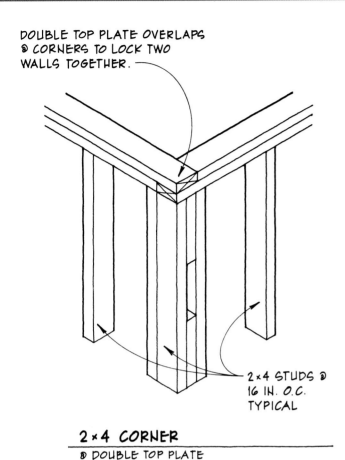

2×4 STUDS @
16 IN. O.C.
TYPICAL

2×4 CORNER
@ DOUBLE TOP PLATE

EXTRA STUD ADDED PERPENDICULAR
TO CORNER STUD PROVIDES NAILING @
INSIDE CORNER & ALLOWS SPACE FOR
4-IN. THICK INSULATION @ CORNER.

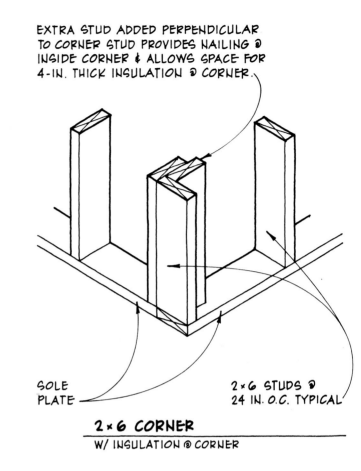

SOLE
PLATE

2×6 STUDS @
24 IN. O.C. TYPICAL

2×6 CORNER
W/ INSULATION @ CORNER

DOUBLE TOP PLATE
OVERLAPS @ CORNERS,
LOCKING TWO WALLS
TOGETHER.

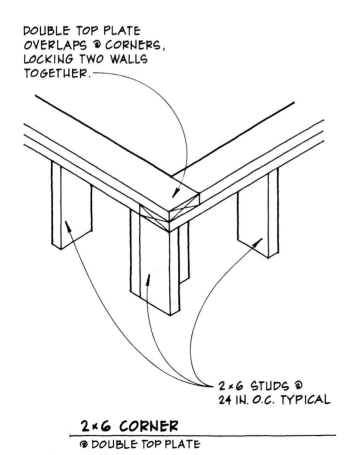

2×6 STUDS @
24 IN. O.C. TYPICAL

2×6 CORNER
@ DOUBLE TOP PLATE

METAL BACKUP CLIPS @ INSIDE CORNERS
OF GYPSUM WALLBOARD ELIMINATE NEED
FOR EXTRA STUD, ALLOWING FOR FULL
THICKNESS OF INSULATION.

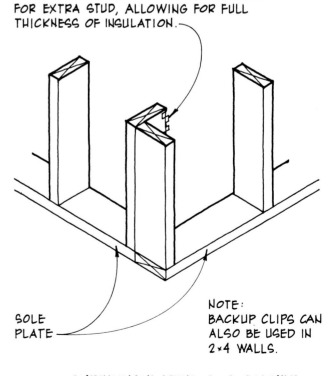

SOLE
PLATE

NOTE:
BACKUP CLIPS CAN
ALSO BE USED IN
2×4 WALLS.

SUPERINSULATED 2×6 CORNER
OUTSIDE CORNER ONLY

Riding Out the Big One

A structural engineer explains what happens to a wood-frame building in an earthquake and how it can resist collapse

by Ralph Gareth Gray

The area affected by the Loma Prieta earthquake, which shook up much of the San Francisco Bay Area in October of 1989, included 1,544 California public schools. Only five of them suffered severe damage. No lives were lost, and there were no injuries in these buildings. Most of them were wood-frame construction very similar to that of a typical custom house, and they were either built from scratch or upgraded following the five sacred principles of earthquake-resistive construction.

There's nothing mysterious about these principles, and I'll discuss them all in this article. To understand them, you must first visualize clearly how a structure carries the loads imposed on it by an earthquake. To put the principles to use in the real world of construction, you might have to make a few simple but critical modifications to standard building practice. Also, you must pay close attention to details, and to get them done right, you'll probably have to be steadfastly persistent. This is a big topic, so in this article I'm going to talk mostly about new construction. In a subsequent piece we'll look at retrofitting existing houses to withstand earthquakes.

Some of what follows is subject to continuing debate among structural engineers, and I'll probably rub some of my colleagues the wrong way. But shall I tell you what I really think, or

dose you with bland, safe consensus? I choose the former.

Sudden shock—In an earthquake, the ground moves violently and chaotically, up and down, side to side, twisting and rocking, with all these motions changing very quickly. Anything on the ground, such as a house, will tend to slide and overturn. Various parts will rattle around on their own and maybe come adrift. Things stacked one on another, like bricks on deteriorating mortar, will tend to slide and overturn independently, and framing members, like beams, may come right off their posts, or the posts off their footings. The connecting links—particularly tension and shear-carrying components—are put to the supreme test in an earthquake. If they are incorrectly designed or constructed, they will cause serious trouble. Thus the first principle: *Tie it together.*

Overlaps and straps—The platform frame holds together better in earthquakes than the balloon frame, in part because a platform frame is tied together by the overlap of the double top plates on the walls at every level. But they must be spliced correctly, with minimum 4-ft. overlaps (top drawing, facing page). While the code requires at least two 16d common nails on each end of the splice, I think four 16d

nails is a better minimum. If more are needed, the drawings should show them. As you add more nails, follow a nailing pattern like the one shown in the drawing to avoid splitting the wood.

Post-to-beam connections are also important, and any measure of reinforcement is better than the following common condition: a couple of toenails through the bottom edge of the beam into the post's end grain. Instead, use steel connectors, such as commercially available post caps. You can make an effective site-built post-to-beam connector using a 2x bolster and a couple of 2x yokes on either side of the post to cradle the beam (bottom drawing, facing page).

Steel column bases are the best way to tie posts to their piers (drawing below, facing page). But even minimal connections, such as short steel Ls secured with drilled-in anchor bolts, or even plumber's tape affixed with concrete nails are better than a couple of toenails.

Parts other than framing members should also be tied together. Use metal straps to secure mechanical equipment, such as the water heater, to the walls. Concrete or clay roofing tiles (and roofing slates) should all be anchored to the roof sheathing with corrosion-resistant fasteners. Unfortunately, there are roof-tile systems, with ICBO approvals, that do not require mechanical fasteners above the first few courses. I shudder to think what might happen to people who are running out the door during a serious shake when the tiles depart the roof.

Stucco can be a fine finish, and it's excellent fire protection, but it has to be tied to the wall. This means that the mesh must stand off the sheathing far enough to key and support the scratch coat, and the nails mustn't pull out of the sheathing because it's decayed or because the nails are too short, rusted or both. In San Francisco's Marina District I saw great sheets of stucco that had peeled right off the walls. If the Marina fire had spread, that stucco wouldn't have protected anything but the paving it was lying on.

Inertia—The sudden movement of the ground under the building and the rapidity of change in the ground's motion set earthquakes apart from other dynamic actions on buildings, such as wind. The sliding, overturning and so on are primarily due to inertia, a physical object's reluctance to be moved, and once moving, its reluctance to change direction or velocity.

The heavier the object, the greater its inertia. The heavier the building, the harder it is to control the forces applied by an earthquake. Thus the second principle: *Keep it light.*

If you're dead set on having a handsome veneer of brick on the sides of your house, remember that the inertial force due to its weight in sudden motion must be channeled successfully through the structure to the ground. That's because our first principle says that we have to tie it together. If you don't tie it, the force will be smaller, but only because the brick will have been thrown off the building, like great lumps of shrapnel. I have clients with expensive earthquake damage to their house that occurred because the brick veneer added to the effective weight of the building, and the structure couldn't handle the increased inertial forces.

Brick chimneys act about the same way as brick veneers, but more so. They are tall and narrow, so the whiplash tension across the mortarbed joints is significant. They are seldom tied to anything, they typically break off in an earthquake and their weight makes them deadly. I try to get people to take down the brick chimney and put in a metal prefab fireplace and double-wall metal flue. If the whole thing can't be removed for some reason, then take it down to the smokeshelf and go up from there with a metal flue. Prefab metal transition pieces are made for precisely this application. Be sure to grout it in so hot gas can't escape.

Tracking the loads—Another key to a building's survival in an earthquake is a strong, stable path for the transmission of inertial forces to the ground. Engineers size the fasteners, shear panels and boundary members (framing elements that take tension and compression loads) that transmit these forces based on a portion (typically around 15%) of the building's weight.

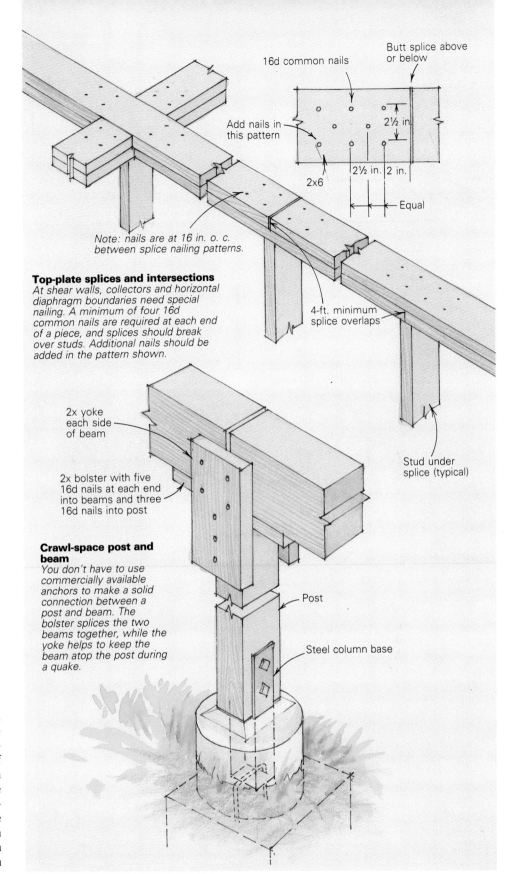

Top-plate splices and intersections
At shear walls, collectors and horizontal diaphragm boundaries need special nailing. A minimum of four 16d common nails are required at each end of a piece, and splices should break over studs. Additional nails should be added in the pattern shown.

Note: nails are at 16 in. o. c. between splice nailing patterns.

16d common nails
Butt splice above or below
Add nails in this pattern
2½ in
2½ in. 2 in.
Equal
2x6
4-ft. minimum splice overlaps
Stud under splice (typical)

2x yoke each side of beam

2x bolster with five 16d nails at each end into beams and three 16d nails into post

Crawl-space post and beam
You don't have to use commercially available anchors to make a solid connection between a post and beam. The bolster splices the two beams together, while the yoke helps to keep the beam atop the post during a quake.

Post

Steel column base

Understanding the forces at work during an earthquake can take some mental gymnastics (see load-path sidebar, next page). But when it comes to designing, building, repairing and retrofitting buildings to resist earthquakes, being able to track the inertial forces through their load paths is an essential skill. Therefore our third principle: *Know the load paths.*

If you can see a plausible path for loads to travel when you study the potential shaking of

a house in each of the principal directions, you are a long way toward an effective earthquake-resisting system, and with this knowledge, you're also in a position to understand some of the more subtle but still important aspects of anti-earthquake construction.

As you can see from the drawings of our little house (see drawings in sidebar), the shapes of its various parts change during the violent shaking of an earthquake. If there's anything in the way, like a tree, the house next door, or even parts within the same structure, it gets hit with a lot of energy—sometimes repeatedly. While this battering dissipates energy, it can cause big chunks to fall off a house and even cause an otherwise sound building to collapse. This leads to our fourth principle: *If you can't tie it together, separate it.*

Obviously we can't separate elements that depend on each other for load transmission, such as a beam from its post. But what about separate portions of a house that might tangle during an earthquake? Consider, for example, a split-level house that steps down a hillside. Assume that the lower part of the house has a big roof abutting a stud wall that carries the weight of another roof (the one over the uphill portion of the house). During an earthquake, the lower roof will try to shake at a different frequency. Thus, the studs might have to be huge—maybe 3x8s at 12 in. o. c.—because in an earthquake, the roof would push and pull against them with great force. In a case like this, it could make sense to separate the lower roof and the walls from the upper wall by 2 in. or so (by about ¼ in. per foot of wall height), which means building a separate studwall to support the lower roof rather than nailing it into the studwall supporting the upper roof. Between the upper wall and the lower roof, you should provide flashings with slip joints so the elements can shake around in all directions without tearing.

A sewer, gas or water pipe passing through a wall and then turning down into the ground is locked into the wall if there isn't space around it at the wall. Leave a ½-in. gap around all sides of such lines, and fill around them with flexible caulk.

Myth of the fantastic optimum—Building structures are analogous to chains, with each element serving as an appropriately sized link to carry out its function. In a perfect world you could design and build a house using the very minimum number of fasteners, the smallest allowable foundation and the lightest structural members. Unfortunately, that's an unsafe and unrealistic approach, because the loads from a strong earthquake are just plain unpredictable. This leads to our fifth (and most important) principle: *Buildings should fail gently.*

Notice that our model building has a shear wall on each side of the floor diaphragms, for a total of four. It's possible to get by with three, provided they are laid out correctly (they should never align to a point, like spokes in a wheel). But even with three, if you lay them out correctly, there's no reserve for construction or design errors, dry rot or future remodeling by klutzes. If one of those three walls fails, for whatever reason, the whole house will fail. This is a plea for redundancy in the structure—the provision of more than just the minimum load path. Put some plywood on other walls (not shown on our drawing), all the way up to the roof, so that if one element fails, the loads flowing through the structure *(Text continued on p. 108)*

Load path

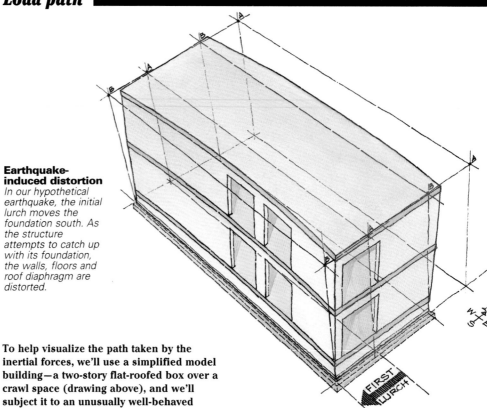

Earthquake-induced distortion
In our hypothetical earthquake, the initial lurch moves the foundation south. As the structure attempts to catch up with its foundation, the walls, floors and roof diaphragm are distorted.

To help visualize the path taken by the inertial forces, we'll use a simplified model building—a two-story flat-roofed box over a crawl space (drawing above), and we'll subject it to an unusually well-behaved earthquake: first a sudden lurch toward the south, followed by a sudden lurch back to the north. Period. We'll start at the bottom of the building and go up, concentrating on primary effects.

The first lurch—You can see where the house starts from by the ghost outlines defined by the A labels on the corners of the roof. During the first lurch, the building would immediately follow the ground motion exactly to the ghost outlines labeled B on the roof corners if it weren't for inertia. The building's inertia causes it to lag behind the ground, as shown by the rendering.

A good foundation is embedded in the ground, so during the first lurch it is carried south along with the ground, taking the jack walls around the crawl space with it because their mudsills are anchor-bolted to the footings. The tops of the walls lag behind a bit, and the east and west walls are changing shape from a rectangle to a parallelogram—a kind of distortion called shear distortion—hence the name shear wall. If they were weak, the east and west walls would just mash over to the north. But they are sheathed with plywood (or diagonal sheathing) and are thus stiff and strong. Unsheathed jack walls are a common weak link in older wood-framed buildings.

The jack walls on the north and south are also sheathed, but can't resist north-south motions at all, for their stiffness and strength work only in an east-west direction. They follow along because their tops are nailed to the first floor, rotating as if on a piano hinge along the sill.

I've shown jack walls here because they're often used on the West Coast to elevate first-floor joists above grade and to level floors on hillsides. The same hinge principle described here, however, also applies to floor assemblies that bear directly on mudsills and stem walls.

The inertial force of the bottom half of the north and south jack walls goes to their footings and the force from the top half goes to the edge of the floor. So for this south lurch, the north and south walls are part of the load, not part of the resistance.

The east and west jack walls drag the first floor (and everything above it) toward the south, but reluctantly. That's because the first floor has its own inertia, plus some contributed by the north and south walls, and any attached partitions or mechanical equipment.

The middle of the floor lags behind its ends, the north and south edges changing from straight to curved. The south edge gets a little shorter and the north edge gets a little longer—a typical sign of bending distortion. Beams and girders act in bending (as well as shear), and that's what the floor is: a big, flat beam that resists lateral forces in a horizontal plane. It's usually called a horizontal or floor diaphragm.

The next layer of the cake, first-story walls plus second-story floor, acts much the same way. So does the top layer—the second-story walls plus the roof.

To summarize from the top down, the roof diaphragm carries its inertial load to the east and west second-story shear walls. The drawing labeled "Distorted horizontal diaphragm" (drawing right) shows a plan view of how the floor might look in mid-lurch. The little arrows represent the inertial loads from the floor itself, and those delivered to the floor from the north and south walls. Bigger arrows at the east and west ends represent the accumulated inertial loads delivered by the east and west shear walls from above. The largest arrows represent the resistance of the shear walls below, and ultimately, the foundation. They are equal to the sum of

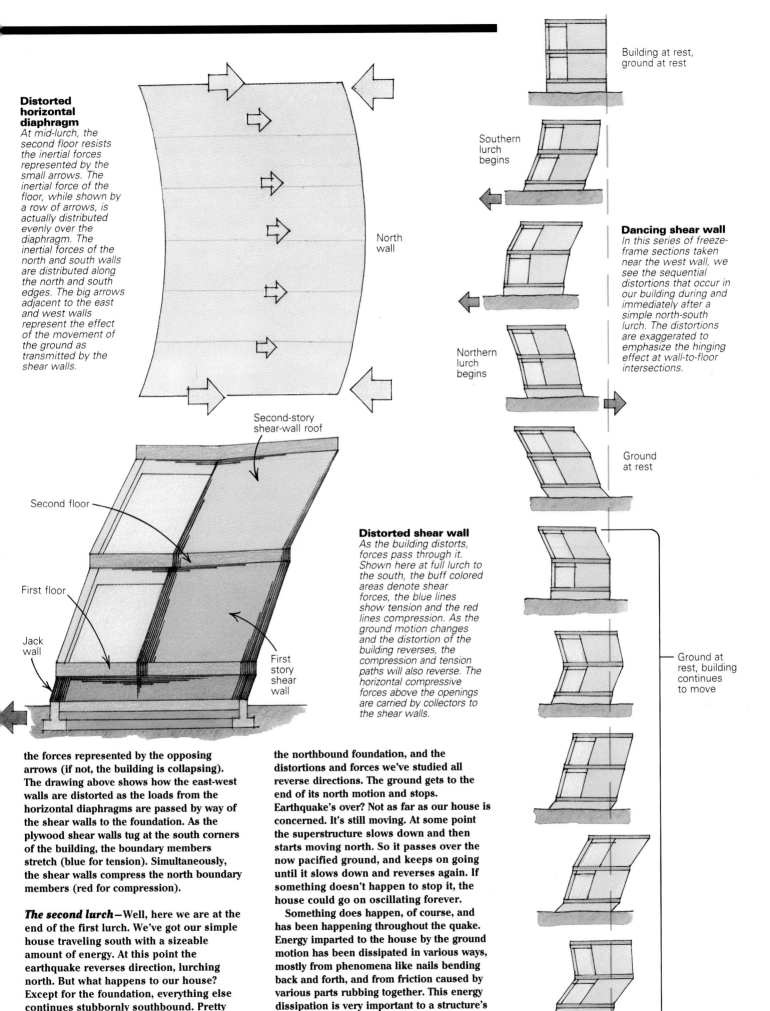

Distorted horizontal diaphragm
At mid-lurch, the second floor resists the inertial forces represented by the small arrows. The inertial force of the floor, while shown by a row of arrows, is actually distributed evenly over the diaphragm. The inertial forces of the north and south walls are distributed along the north and south edges. The big arrows adjacent to the east and west walls represent the effect of the movement of the ground as transmitted by the shear walls.

North wall

Building at rest, ground at rest

Southern lurch begins

Dancing shear wall
In this series of freeze-frame sections taken near the west wall, we see the sequential distortions that occur in our building during and immediately after a simple north-south lurch. The distortions are exaggerated to emphasize the hinging effect at wall-to-floor intersections.

Northern lurch begins

Ground at rest

Second-story shear-wall roof

Second floor

First floor

Jack wall

First story shear wall

Distorted shear wall
As the building distorts, forces pass through it. Shown here at full lurch to the south, the buff colored areas denote shear forces, the blue lines show tension and the red lines compression. As the ground motion changes and the distortion of the building reverses, the compression and tension paths will also reverse. The horizontal compressive forces above the openings are carried by collectors to the shear walls.

Ground at rest, building continues to move

the forces represented by the opposing arrows (if not, the building is collapsing). The drawing above shows how the east-west walls are distorted as the loads from the horizontal diaphragms are passed by way of the shear walls to the foundation. As the plywood shear walls tug at the south corners of the building, the boundary members stretch (blue for tension). Simultaneously, the shear walls compress the north boundary members (red for compression).

The second lurch—Well, here we are at the end of the first lurch. We've got our simple house traveling south with a sizeable amount of energy. At this point the earthquake reverses direction, lurching north. But what happens to our house? Except for the foundation, everything else continues stubbornly southbound. Pretty soon the southbound superstructure passes

the northbound foundation, and the distortions and forces we've studied all reverse directions. The ground gets to the end of its north motion and stops. Earthquake's over? Not as far as our house is concerned. It's still moving. At some point the superstructure slows down and then starts moving north. So it passes over the now pacified ground, and keeps on going until it slows down and reverses again. If something doesn't happen to stop it, the house could go on oscillating forever.

Something does happen, of course, and has been happening throughout the quake. Energy imparted to the house by the ground motion has been dissipated in various ways, mostly from phenomena like nails bending back and forth, and from friction caused by various parts rubbing together. This energy dissipation is very important to a structure's survival in an earthquake.   —*R. G. G.*

All other drawings: Malcolm Wells

will have an alternate path. The house might sag or show cracks but won't collapse, killing someone. Such precautions will cost a few extra dollars, but that's cheap insurance.

The horizontal diaphragm—During an earthquake, the floor and roof diaphragms undergo shear and bending. The subfloor sheathing carries the shear, like the web of an I-beam, while the boundary members (rim joists and top plates) carry the tension and compression due to bending like I-beam flanges.

To withstand and transfer the shear loads, plywood sheets have to be spliced together to prevent adjacent edges from sliding past or over each other. Plywood sheet edges should be butted and nailed to joists in one direction, and to solid blocking or rim joists in the other. Butted on the centerline of a 2x joist, you've got only ¾-in. bearing for each piece, so the nail has to be ⅜ in. from the edge. The edge-nailing called for by code can be as close as 3 in. o. c. This layout works, but there is no margin for error. Layouts must be accurate, and the nailing has to be done with care to avoid shiners and split joists or blocking.

Plywood or diagonal board sheathing is edge-nailed to the rim joists or blocking on all four sides of the diaphragm. They in turn must be connected to the top plates below, which serve as chords (like I-beam flanges), carrying tension or compression.

The edge joists and top plates compose the boundary members. Walls are often longer than the lumber available, so top plates or edge joists must be spliced for the tension and compression. If they've been severed, boundary members must be spliced (plumbers are especially good at finding and disabling the most critical diaphragm chords, usually because designers have given them no alternative). If for instance, a vent stack has been let into the side of a top plate, the load in the top plate becomes eccentric, magnifying the stress on it during an earthquake. This may snap it. Custom-made splints of ¼-in. steel angle, for example, may be required to fix this.

Large openings, like stairwells, need boundaries around them. Put blocking and strapping perpendicular to the joists across several joist spaces to compensate for the local increase in shear and bending due to the hole (drawing right). The same kind of detailing is needed at inside corners of L-shaped or more complicated diaphragms.

Shear walls and collectors—Just as the horizontal diaphragm is a big, flat beam that resists lateral forces in its horizontal plane, the shear wall is a big, flat vertical cantilever beam that resists lateral forces in its vertical plane. Again, the plywood or diagonal sheathing carries the shear, and the boundary members—stud corners or end posts—carry the bending tension and compression. Shear walls tend to be harder to engineer than floor and roof diaphragms, in part because they're smaller (one does need doors and windows). Also, loads accumulate from the top down, so the

loads tend to be just plain bigger. All plywood edges, horizontal and vertical, should bear on, and be nailed to, studs, plates or blocking.

The connection between a shear wall and its foundation typically serves two functions: the transfer of shear forces delivered by the wall to the ground by way of the foundation; and the transfer of overturning forces (called uplift) to the foundation. Anchor bolts take care of the shear at the foundation level—the larger the shear force, the larger or more closely spaced the bolts. Tiedown anchors, bolts, straps or other devices resist uplift.

The shear forces from the roof boundary members are transferred to the top of the shear wall in several ways. They pass by way of nails that slant from the edge joist or blocking into the top plate, or by flat blocking between the joists nailed in turn to the top plate (the flat blocks can also be used for drywall backing). Another method is to run the wall plywood an inch or so up onto the blocking or edge joist (drawing facing page). Here, the heavy edge-nailing schedule for the wall will be used, and another line of nails will be embedded into the center of the top plate. This detail has many forms. Note that the plywood does not run all the way to the top of the joists and blocks. That's because they will shrink, while the plywood does not, causing humps in the walls and sometimes splitting the boundary members. At the joists between floors, the bottom edge of the plywood should not get too close to the top edge of the wall plywood below, because the cross-grain shrinkage of the floor or contact during the quake will strip the plywood off, split the floor members, or both. Leave about a 1-in. gap between them—enough to account for shrinkage.

At the base of each wall, shear is trans-

ferred from the plywood to the sole plate by nails, and then from the sole plate to the floor plywood. Special nailing schedules sometimes apply to this connection.

Shear walls in the lower stories resist accumulated shear, uplift and compression from the walls stacked above. Think of it all as one wall, continuous from foundation to roof, spliced at intersecting floors.

As shear forces move through walls, they have to take a path around openings for windows and doors. The forces are concentrated in the boundary members over the openings before they can be dumped into the shear wall. The framing members that handle this task are called collectors, or drag struts. Often the top plates above the headers serve this function—another reason why the top-plate splices are important. Sometimes beams are used as collectors. This is a tricky little item that is very important and often overlooked, so be aware when there are long openings in the plane of a shear wall, particularly if the roof or floor above is flush-framed. Like top plates, collectors are targets cherished by plumbers.

Fasteners—No structural element is better than its connections, and no connection is better than the fasteners. The common nail is a magnificent device for resisting earthquakes. Tests I helped conduct at the University of California's structural research facility showed that properly nailed plywood shear walls have an amazing capacity to resist earthquakes and dissipate energy owing to the ability of the nails to flex back and forth repeatedly without breaking.

But, the plywood walls secured by *overdriven* nails (nails that penetrated the plywood beyond the first veneer) failed suddenly in our tests, and at loads far below those carried by

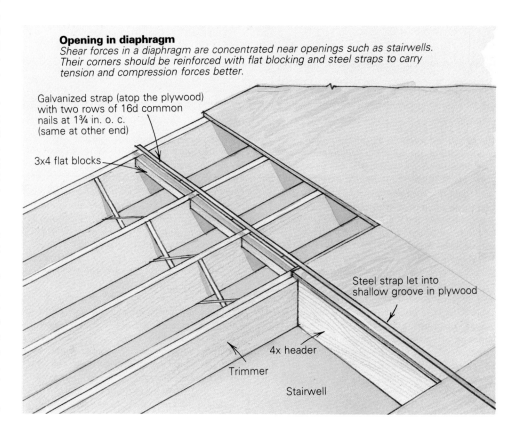

Opening in diaphragm
Shear forces in a diaphragm are concentrated near openings such as stairwells. Their corners should be reinforced with flat blocking and steel straps to carry tension and compression forces better.

Galvanized strap (atop the plywood) with two rows of 16d common nails at 1¾ in. o. c. (same at other end)

3x4 flat blocks

Steel strap let into shallow groove in plywood

4x header

Trimmer

Stairwell

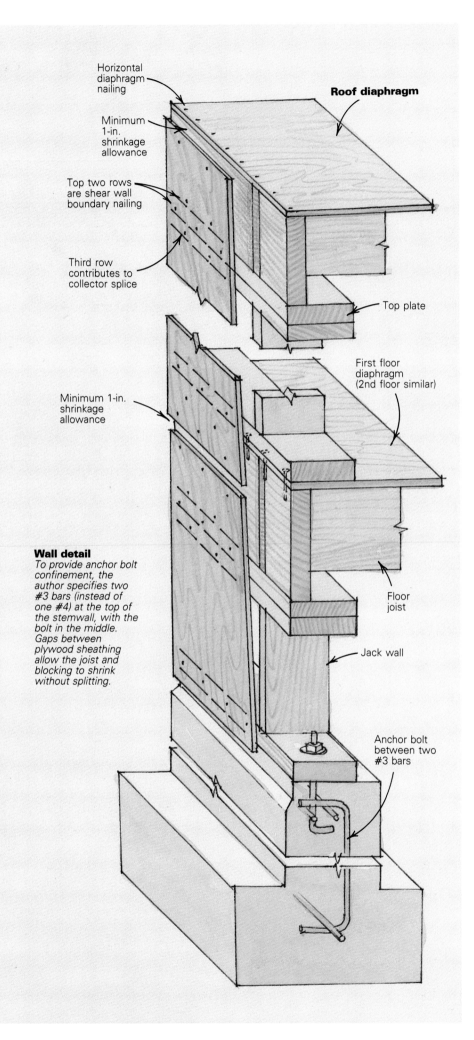

Horizontal
diaphragm
nailing

Roof diaphragm

Minimum
1-in.
shrinkage
allowance

Top two rows
are shear wall
boundary nailing

Third row
contributes to
collector splice

Top plate

First floor
diaphragm
(2nd floor similar)

Minimum 1-in.
shrinkage
allowance

Wall detail
To provide anchor bolt
confinement, the
author specifies two
#3 bars (instead of
one #4) at the top of
the stemwall, with the
bolt in the middle.
Gaps between
plywood sheathing
allow the joist and
blocking to shrink
without splitting.

Floor
joist

Jack wall

Anchor bolt
between two
#3 bars

correctly nailed plywood panels. Overdriven nails are typically installed by careless nailgun operators. If the gun sets the nails erratically, back off on the pressure, let them stand a little proud and drive them flush by hand.

In our tests, stapled plywood shear walls performed pretty well, but they weren't as strong as the nailed walls. Staples, being cold-worked, may be susceptible to brittle-failure, and being thin, subject to corrosion. This is an unhappy combination. So waterproofing the wall is more important when using staples than when using nails. Overdriven staples reduced strength, but not so badly as overdriven nails. Drywall, great for fireproofing and finish, should not be used for resisting earthquakes. As the walls flex, the nails just excavate little slots in the drywall. Bugle-head drywall screws, annular-shank or threaded nails, or regular screws are cold-worked, so they can't stand the repeated reversal and extreme deformation that common nails can. Don't use them for shear walls, unless you're certain they are annealed to the performance level of a common nail.

Lag bolts are fine for connections where loads are concentrated. But to work properly, the hole has to be drilled twice—once for shank and once for the threads—then lubricated with paraffin wax and the bolt turned in. No hammering allowed.

Another fastener that's subject to improper installation is the anchor bolt. It carries shear loads from the mudsill into the footing, or would if it weren't too close to the edge of the footing or outside the line of any rebar that might prevent the concrete from spalling during a quake. An equally useless method for installing anchor bolts is to stab them into the concrete after the pour has initially set—a disgusting practice that guarantees the bolt shank will be "anchored" in a cone of laitance (weak and crumbly concrete). Prior to the pour, anchor bolts should be wired inside a double run of rebar at the top of the stemwall (drawing left). The bolts' threads should extend far enough above the finished level of the concrete to accommodate the mudsill, the washer and the nut.

Moisture and shrinkage—Don't expect anything to work structurally when you really need it if it's decayed or being digested by termites. Make sure you've protected your work. Decay is the more insidious of the two. At the point when a specialist can only marginally detect decay under a powerful microscope, *80%* of the wood's shock resistance has vanished.

Strap ties used as tiedowns between floors will buckle as the joists dry, unless they're installed after the joists have shrunk. For the same reason, bolted tiedowns need to have their nuts tightened just before the walls are closed in. □

Ralph Gareth Gray is an architect and structural engineer living in Berkeley, Calif. He has designed wood-frame buildings for more than 30 years and served on code advisory committees for the Structural Engineers' Association of California and the American Institute of Architects.

Posts can be assembled in place. The author screws a track onto a pair of studs to complete a built-up post. To his right, a pair of cantilevered headers meets at an outside corner. The headers are stuffed with insulation and anchored to one another with steel straps.

Framing With Steel for the First Time

A builder accustomed to wood framing tells about the differences between wood and steel construction

by Robert McCullough

Until two years ago, I'd spent my construction career building houses of wood. Then I got a tempting offer. My father-in-law, architect Berle Pilsk, wanted me to build his new house at Sea Ranch, California. The thought of spending a year living on the northern California coast and building a house in a redwood grove was enticing. The only hitch: The house was to be of steel.

Berle is convinced that steel-frame houses represent the next advance in residential construction. A steel frame doesn't rot, insects won't eat it, it won't burn, it resists earthquakes well, the parts are of uniform dimension, and—perhaps best of all—at this writing the cost of steel components is substantially less than comparable wood components on the northern California coast.

But I was skeptical. I like working with wood. It smells good, and it's easy to cut, shape and fasten. So before I took on the job, I spent a lot of time with Berle, going over the hypothetical construction of the house. I found a couple of steel houses that were under construction and studied them. I learned that a steel-frame house is essentially the same as a stick-frame house—the parts are just made of different materials. But there are some significant differences that you should be aware of if you're familiar with wood construction and if you're contemplating your first steel-framing project. For information on specific construction details, get in touch with the American Iron and Steel Institute (800-797-8335). The group's technical-data brochure, *Low-Rise Residential Construction Details* ($20), is filled with easy-to-decipher assembly drawings. What's more, that phone number will hook you up with their technical-advice hot line.

A 2x6 is really 6 in. wide—Steel-framing members consist of two basic components that are C-shaped in section, with a couple of important differences. Studs, joists and rafters are made with members that have flanges folded inward about ¼ in. at their open corners (top photo, facing page). These folds, while almost insignificant dimensionally, add stiffness to the members and make it easy for a stud to stand vertically. Studs come with prepunched holes in their webs for electrical conduits. Joists and rafters can have solid webs or be prepunched.

The other basic components are tracks. They have solid webs and lack the folded corners on the legs. Tracks work both as sill plates and top plates, and as part of posts or headers when combined with studs. Tracks also can be used for blocking and backing.

Like wood components, studs, joists, rafters and tracks come in various dimensions that differ in regular increments. But unlike wood, the dimensions are a net figure. In other words, 6 in. means 6 in. Studs are measured from their outside dimensions, and tracks are measured from the inside of their legs. So a 6-in. stud fits snugly into the track as shown in the photo. Self-tapping screws driven through the legs of the track and the legs of the stud hold the assembly together.

Steel-framing members come in many different sizes and gauges. We were able to order and receive floor joists and rafters in excess of 30 ft.

They were all the same length, and they were all perfectly straight. There is no sorting through the pile looking for the "perfect timber," although I must confess to picking up a stud from time to time and absentmindedly sighting down its length. Old habits are hard to break.

Steel components are surprisingly light for their strength. Without much difficulty, one person can carry a 30-ft. rafter. Studs are shipped nested together in bundles of ten. One person can pick up and carry a bundle of 10-ft. studs with relative ease. They don't take up much space, and it doesn't matter how long they sit. They will never dry, warp, split, rot or burn.

I had no problem with getting my hands on the stuff. Deliveries were prompt and on schedule, and our supplier needed a minimum of advance notice. The steel usually comes on a large flatbed with many small bundles banded together into several large and heavy bundles. The driver can't just drop the load as with lumber because steel distorts and bends when it hits the ground. A forklift is the best tool for unloading the material. In addition to making quick work of the job, a forklift allows you to keep large bundles together until you need the materials.

By the way, our steel came coated with a water-based lubricant that made it tricky to handle. Gloves are required. Once the steel sits out in the weather, it loses the slippery quality, but gloves are always a good idea. A word to the wise: When ordering, order carefully because you can't run down to the lumberyard and pick up a few more sticks—yet.

One of the attractive things about framing in steel is its amazing strength. You can increase the strength, and hence the load-bearing capabilities of the member, simply by increasing the thickness, or gauge, of the metal. You don't have to increase the dimensions of the member. When you increase the thickness of a steel member, the gauge number gets smaller. For example, a 12-ga. stud is beefier than a 20-ga. stud.

Screws hold it all together—The numbering system works the opposite way for the screws that hold a steel frame together. A #8 screw is much smaller and does not provide as much strength as a #14 screw.

Screws have either Phillips-head recesses, square-drive recesses or hexagonal heads. We used all three. Phillips and square-drive screws work best for thinner materials. When we had to screw through anything thicker than 14 ga., we used hex-head screws because it's tough to deform a hexagonal screw head, even under the constant, intense torque it takes to get a screw into thick sheet metal. By the way, it is important—almost mandatory—to keep a hex-head screw perpendicular to the material when you're driving it home. If you angle a hex-head screw as it is driven, it will simply spin out of the drive socket. A Phillips-head screw, on the other hand, can be angled into the work and still be driven, which allows you to get in some tight places. Pan-head screws are designed to be used in places that will be covered with drywall, where the fatter hex-head screws would cause a bulge.

Components and fasteners

Framing members

The parts are folded sheet metal. The basic components in a steel-frame house are C-shaped in section. On the left, a stud nests in a track. Note how the legs of the stud curl inward, strengthening the form. The prepunched hole in the stud is for wiring. In the middle, a built-up post is composed of two studs and two tracks. The part on the right is a hat channel. It's typically used for purlins.

Fasteners

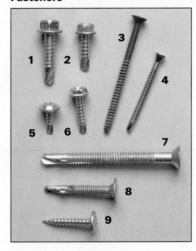

Choose the right screw for the job. Most steel-framing screws have self-drilling tips that eliminate the need for drilling a pilot hole (see photo below). Phillips-head and square-drive screws are good for 20-ga. and thinner metal. Hex-head screws are best for thicker metal. Use the wing-tip screws for attaching wood to metal.

1. #14, 1¼-in. hex washer-head screw.
2. #12, 1-in. hex washer-head screw.
3. 2¼-in. Phillips bugle-head screw.
4. 1⅝-in. square-drive flathead trim screw.
5. #8, ½-in. Phillips pan-head screw.
6. #10, 1-in. hex washer-head screw.
7. #12, 2½-in. Phillips flat countersunk-head screw with wings.
8. #10, 1½-in. Phillips flat wafer-head screw with wings.
9. #8, 1-in. Phillips pan-head screw.

Screw tips

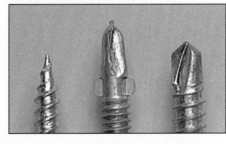

Self-tapping and self-drilling tips. You can drive a self-tapping screw (the pointy tip on the left) into the thinner steel components. If the steel is thicker than 20-ga., however, use a fastener with the self-drilling-type tip on the right. The screw in the center, which is used to affix wood to metal, has a self-drilling tip with little wings between the tip and the threads. The wings carve the hole in wood, then break off when they encounter steel.

Screwing steel-framing components to one another is not difficult. It is, however, time-consuming. For example, the basic screw for anchoring a stud to a track on our job was the #8, ½-in. pan-head self-drilling screw (middle photo above). Over the course of building this house, we went through 20,000 of these screws. Let's say we dropped 1,000 on the ground. This leaves 19,000 screws. On the average, it takes about 30 seconds to drive one of these screws. Convert 9,500 minutes into workdays, and you get almost 20 eight-hour days.

A new set of tools—You can get material precut to specific lengths, but we decided to cut the components on site because of the intricacies of the house. The cut-off saw (photo p. 112) is the best tool for the job. After researching the field, I concluded that all the saws are pretty much the same and ended up buying the Makita.

Although the saws are similar, there is a big difference in the blades. After trying every available metal-cutting blade we could find, we concluded that the ones made by Norton lasted the longest (Norton Co., 1 New Bond St., Worcester, Mass.

the foundation because the steel track is not as forgiving as a wood mudsill atop a layer of sill sealer. We used traditional ⅝-in. foundation bolts for the typical stretch of foundation. Where the engineer called for heavier connections to the foundation, we used ¼-in. steel plates affixed to the foundation with ⅞-in. bolts. The plates were then connected to the frame by way of diagonal braces made of 16-ga. steel.

One of our first problems to solve was how to cut the holes in the track for the foundation bolts. At first I used a bimetal hole saw, but I found this to be difficult and sometimes painful. A hole saw has a tendency to bind in the kerf and then jerk your wrist. I finally found a stepped drill bit, which is cone-shaped and starts small enough not to need a pilot hole but ends up big enough to create a 1-in. hole. This tool also came in handy when it was time to install plumbing and electrical. Although the studs came prepunched, we sometimes needed to make holes in built-up posts and headers to run wires and pipes.

Accurate layout of framing members is important no matter what the material, but with steel layout accuracy takes on greater importance. Rafters must be over either a post or a double stud. If the house is more than one story tall, posts and studs must bear on a plate that is supported either by a joist or a short piece of stud called a web stiffener placed in the joist track directly under the post or stud (top photo, facing page). These loads must be transferred to the foundation by corresponding posts and studs in the lower floors. So when we were laying out first-floor walls, we were also laying out roof framing.

Frame the walls in place—A carpenter typically lays out the parts of a wood wall on the subfloor, frames the wall in the horizontal position and then lifts it into place. A steel-frame wall requires that the screws be driven from both sides of the track into the legs of the stud. So we framed the walls, which have their studs on 24-in. centers, in place rather than flip the partially completed wall to gain access to the other side (for another approach, see sidebar p. 114-115).

After marking the position of the stud (pencils worked fine), we placed a stud in the track. Then we clamped it in place with the Vise-Grip C-clamp and tapped the stud firmly downward to make sure it was seated correctly. This is an important step. The American Iron and Steel Institute recommends a gap of no more than ¹⁄₁₆ in. between a stud and its track. If the gap is larger, the screws end up as the bearing points.

After tapping the stud down, we ran the appropriate screw through the track and into the leg of the stud. Then we clamped the other side of the stud and screwed it to the track. For bearing walls, we found it easiest to install all the studs and then come back and add the top track. Walls that intersect at corners are attached with screws through the abutting studs and by straps that wrap around the tracks at the top of the wall.

Partition walls were easier to build by positioning the top and bottom tracks and then filling in the studs. This step should be done only with nonbearing walls because it's tough to make sure there aren't gaps larger than ¹⁄₁₆ in. between

Take the studs to the saw. A cut-off saw with a composition blade is the standard method of cutting steel-framing components. Cutting sheet metal this way demands adequate safety protection.

01606; 800-543-4335). On a day of cutting, it's easy to go through a couple of 14-in. sawblades.

Cutting metal tracks and studs is a noisy job that produces steel slivers that can get in your eyes and clouds of acrid smoke that can sting your lungs. Protect yourself with earmuffs or earplugs, eye goggles and a respirator. Also, have a first-aid kit with plenty of Band-Aids, antibiotic ointment and eye wash. It is important to wear the eye protection because a metal sliver is nothing like a wood one. A steel sliver tends to stay in the eye, where it immediately starts to rust.

We used Porter-Cable screwguns for assembling the components and a Bosch battery-powered drill with a long extension to reach the nearly inaccessible places that needed screws (a cordless drill seems to work better than a standard drill if you can't hold the tool perpendicular to the work).

I never did find a store-bought hex-head driver for the #10s that would magnetically hold the screw. The head depth of the #10 is shallower than the #12, so the magnet doesn't quite engage the head of the screw. We solved the problem by adding a thin magnet to the driver socket.

Here's the lineup of tools for steel framing that I carry in my belt: two large Vise-Grip C-clamps, spare Phillips and hex-head screwdrivers, a stringline, plumb bob, magnetic torpedo level, end nips for changing the driver bits in my screwgun and removing backed-out screws, a utility chisel, three pairs of metal snips (straight, right hand and left hand), a 4-in-1 screwdriver, a utility knife, a tape measure, a speed square, my cat's paw and, finally, my hammer, for balance if nothing else. I'd say the whole kit weighs about 25 lb.

The layout has to be impeccable—A steel house is put together in small pieces, just like the typical platform-frame house of wood. The foundations are the same for wood or steel, but with steel it is important to trowel smooth the top of

The view from the crawlspace. Joists at 4 ft. o. c. carry the corrugated metal deck of the main floor. Each joist is carried by a pair of studs. At the pony wall in the background, web stiffeners midway between the joists carry the loads from the walls above. Note how the joists are blocked at the rim tracks by short sections of steel studs and at midspan by blocking made of steel studs.

studs and tracks. Because we were building on a sloping site, we had to erect pony walls (sometimes called cripple walls or jack walls) of various heights to establish a level top plate. Then we were ready to lay out our floor. Incidentally, we sprayed the cut ends of any framing members in the crawlspace with some zinc-rich Rustoleum paint.

Rolling out steel joists goes really fast. First, we installed a rim track (a 20-ga. track stood on end) on top of the pony walls and then screwed the joists to the track on 48-in. centers. Halfway between the joists we installed web stiffeners in the track to pick up the load of the studs to come.

We also installed web stiffeners adjacent to each joist, screwing into the side of the joist as well as into the inside of the rim track. The stiffeners are analogous to blocking in this application, supporting the sides of the joists and adding rigidity to the rim track.

A metal deck supports a slab floor—At this point most houses would be ready for a subfloor. But our project has a radiant-slab concrete floor poured over a metal deck. So our next step was to install a 20-ga. metal angle around the inside perimeter of the walls (photo right). Called a closure strip, this metal angle acts as a flange that supports the metal deck at the edges while simultaneously working as a screed for the concrete and a strap tie at the corners to hold the wall tracks together.

Once the closure strip was installed, we laid the 3-ft. by 12-ft. pieces of 20-ga. decking perpendicular to the joists. The corrugations in the deck-

This floor deck hangs on a closure strip. A metal angle called a closure strip supports the edges of the corrugated floor decking and acts as a screed to control the concrete's thickness. In this photo, you can see the closure strip at the outside corner, just inside the overlapping tracks that secure the intersecting walls.

ing nest inside one another, and they overlap one corrugation at their edges. This overlap has to be crimped by a tool that looks like a big pair of bolt cutters. Called button-punching, this operation goes quickly once you have the tool. It took an hour to do the crimps and two days to find and fetch the tool from the rental yard.

If the sections of metal decking, or pans, needed to be cut to length, we used a circular saw fitted with an 8½-in. Norton blade. We affixed the decking to the joists and the closure strips with screws. Screwing the pans to the floor joists really tied the whole platform together, giving us a rigid deck for the main floor of the house. By the way, the light coming off the deck was blinding. We all had to break out the tinted safety glasses when the sun cleared the trees.

If you have occasion to install a metal-pan deck, here are a few things to consider. First, the corrugations in the deck make it tough to keep clean. Debris can be swept into piles along the flutes, but you can't get a dustpan in there. Make sure to have a vacuum on hand. Rain on an unswept metal deck makes a real mess that's tough to clean up. Also, a metal deck isn't much fun to walk on. In our case, the slab was going to be stained and polished to become the finished floor. So we waited until the end of the project to pour the slab. If the slab were to be covered with some kind of finish, say tiles or slate, I certainly would have poured it as soon as possible.

Making headers and posts—Once we had a platform to work on, we made a cut table and mounted the cut-off saw on it. If you build a steel-by Steven Jacobs

During my 16 years in the building business, I've worked as designer and inspector on jobs that ranged from office buildings to a wind tunnel for NASA. That experience prepared me well for my current career, which is designing and supervising the assembly of steel houses on the Olympic Peninsula in Washington state. In the past three years, I've detailed plans for more than 50 houses and been the general contractor for a half-dozen homes.

I like framing with light-gauge steel because for about the same price as wood, you end up with a more structurally sound, termite-proof, rot-proof, rodent-resistant, straight and true structure. What's more, instead of burning waste or paying to dump it, you can recycle nearly all the leftovers. But working with steel, you must be fastidious at every stage of construction. If you just slap a house together, steel framing will be your worst nightmare. I strongly suggest that builders who are new to steel but familiar with wood construction hire an experienced steel framer to work alongside the crew during its first steel project.

Buy extra, and start out dead level—We order approximately 90% of our steel to be delivered cut to length. Track (wall plates) is best ordered in standard 10-ft. lengths. I always order at least 10% more track, and 50 to 100 extra structural wall studs. We use standard stud lengths just as in wood framing, but with steel the studs are 97 in. long. I keep a small stock of the most common studs and joists so that I always have a few extra available. I find that a magnetic level (at least 4 ft. long) is essential, and a 50-ft. to 100-ft. long water level helps immensely in leveling the top plates of the walls.

A flat, level foundation is critical when framing with steel. We make that condition clear to our concrete subs, and we retire and rehire them accordingly.

When the steel is delivered to the site, we lay out the components in an orderly fashion. First we assemble all the headers and beams required to erect the first floor, and we pack them full of fiberglass insulation. Then we lay out the bottom tracks that are to be connected directly to the concrete foundation. We put these tracks atop a ⅛-in. thick, 6-in. wide strip of polystyrene sill sealer. This step prevents electrolysis between the concrete and the steel. At this point we roll out the floor joists and sheathe the floor with plywood. We use #6 Grabber brand self-drilling bugle-head screws to affix the plywood, along with construction adhesive glue.

We lay out all our studs 24 in. o. c., and we frame and sheathe our walls while they are lying on the subfloor or slab. We check every bottom-to-top track dimension at each stud as we assemble the wall, and we clamp a long floor joist to the bottom track to keep it straight as we assemble the wall. Then we check the wall for squareness.

At this point only the two end studs are screwed to their tracks. All others are being held with Vise-Grip clamps. After the wall is squared, we screw off the rest of the connections and sheathe the wall with ½-in. CDX plywood. The plywood sheeting provides shear value as well as acts as a thermal barrier. The wall is now

frame house, I strongly suggest you do the same. Instead of taking the saw to the woodpile, you bring the steel to the cut table.

Posts are made up of two studs and two lengths of track. All our posts were made of 20-ga. material. The studs are placed vertically, with their legs pointing at one another, in the bottom track of the wall. The distance between the studs is equal to the inside dimension of the track that will unite them. I found it easier to install the top track of the wall, and then come back and screw the tracks onto the studs to complete the posts (photo p. 110). Before screwing the tracks onto their studs, stuff the post with insulation. You can't do it after the pieces are assembled.

A header is made the same way as a post, but it's easier to build the header on the ground and then lift it into place rather than to try to build the thing in the wall. The headers we made were 10 in. by 3⅝ in. They are remarkably strong. Because of the strength of the steel, Berle designed a number of corner windows that required no center posts. The roof loads are carried by impressively cantilevered headers.

The rafters don't need bird's mouths—Unlike most wood rafters, a metal rafter doesn't require a bird's mouth because the rafters are held in place atop the wall by a metal angle clip. The angle clip is placed directly over a double stud or post and then screwed to the side of the rafter so that the rafter's load is transferred down through the clip.

Our rafters were attached to our ridge beam with the same type of angle clip. Because of this connection, it was only necessary to cut our rafters to within ¼ in. of the ridge beam. There was no need for an exact measurement between

Purlins anchor the roof deck. Purlins made of steel hat channel run parallel with the ridge on 24-in. centers to make a regular layout for affixing the plywood decking to the roof.

the wall plate and the ridge beam, so this portion of the project went quickly. At the eave line, a track that's screwed to the ends of the rafters acts as blocking.

The spacing of our rafters is irregular, so to gain a consistent layout for the plywood decking, we screwed purlins made of 20-ga. hat channel to the rafters (photo above) at 24 in. o. c. The purlins provide support for sheathing and create a ventilation channel above the insulation.

Wood sheathing and nailers for shingles and windows—Even though the skeleton of this house is steel right down to the foundation, we still found it convenient to skin it with plywood for the roof deck and for the shingled exterior walls. We used #8, 1-in. self-tapping pan-head screws to affix the plywood decking to the 20-ga. purlins. We used self-tapping screws instead of self-drilling screws on the roof deck because it's easier to get force behind the drill when you're pushing down on it.

We used our nail guns to affix the plywood wall sheathing to 2x4 nailers. Attaching the nailers to the steel studs gave us a chance to use yet another kind of specialized screw: a self-drilling screw with wings (bottom photo, p. 111). This screw has a pair of cutters above its self-tapping tip that bore a hole in the wood larger than the diameter of the threaded shank. This larger hole keeps the threads from engaging the 2x4 when the screw spins without penetrating the metal as the tip cuts a hole in the stud. With an ordinary self-drilling screw, the 2x4 climbs the shank as the screw spins. Once the stud is penetrated, the wings break, and the threads drive into the metal.

A magnetic drill press helps with the big holes—In some places we were able to reduce the number of screws required by using larger fasteners. For example, the plans called for ¼-in. self-drilling screws to tie the 16-ga. X-braces to the steel-plate hold-downs. The hold-downs were ¼ in. thick—too much to expect a self-drilling screw to penetrate. Because we had to predrill the holes anyway, we talked it over with the engineer and redesigned the hold-down using ½-in. machine bolts. But how could we drill the holes efficiently? The answer turned out to be a ready to stand and be temporarily braced. We do not screw off the interior-side connections until all the bearing walls are completed and in place. This step allows us to fine-tune alignment and height of all the walls all at once, minimizing our top-plate height variations while maintaining our ⅛-in. allowable tolerance.

We have framed roofs both with site-built steel trusses and with factory-built wood trusses. And we've learned that complicated roofs, low-pitch roofs and most hip roofs can be built with wood for less cost than steel. Because wood-truss assemblies are standardized and are factory built, they are less expensive to buy and take less labor to install. But if you want the advantages of steel framing, almost all roofs can be framed with steel if you have the time and the budget.

Use commercial subs; they're used to steel—Here on the Olympic Peninsula, there are quite a few commercial contractors willing to work on steel-framed residential projects. So we have found it fairly easy to find plumbers, electricians

and mechanical contractors who are willing and able to do our projects at a competitive price.

Insulating steel-frame homes is no big deal. I advise our clients to use the Blown in Blanket System (Ark-Seal, 2190 S. Kalamath, Denver, Colo. 80223; 800-525-8992). It's a fiberglass insulation blown into wall cavities, filling voids in studs, around wiring and electrical boxes, and around plumbing.

The fibers are laced with a glue that holds the insulation in place, preventing settling. Installers use hot glue to stick nylon mesh to the studs. The nozzle that delivers the insulation is then poked through small holes cut in the mesh.

We don't use a plastic vapor retarder between the studs and the drywall. Instead, we finish the drywall with vapor-barrier paint, which limits the transfer of moisture through the walls. If you choose a plastic vapor retarder, however, you can adhere the plastic to the studs with hot glue.

If you want to get your feet wet without risking much, try using steel studs for your interior

nonbearing walls. Steel studs will save you money on material costs, and they will give your framers a chance to get used to the material.

Designers can learn more about detailing steel houses by contacting the American Iron and Steel Institute (800-797-8335). Ask for their handbook RG-930, *Residential Steel Framing Manual for Architects, Engineers and Builders* ($40). Finally, steel manufacturers all have span tables and engineering-specification books free for the asking. You should be aware that some steel manufacturers' span tables assume that you are using web stiffeners at ends and at centers of joist spans, and some span tables don't.

—Steven Jacobs is vice president of Cottage Steel Industries in Port Orchard, Washington.

If you just slap a house together, steel framing will be your worst nightmare.

Milwaukee Electromagnetic Portable Drill Press (photo right), which we rented for the occasion. Flip the switch, and the baseplate on this baby grabs hold of the workpiece with the grip of death. Using this tool made it quick and easy to drill the 110 holes for the braces.

Casings, drywall and baseboards—Unlike the windows, the door openings weren't wrapped with wood. Instead, I screwed a track onto the studs on each side of a door opening, and then screwed the jambs of our prehung doors to the tracks. I used yet another kind of screw in this instance—a square-drive trim screw with a self-drilling point (center photo, p. 111). I used these screws because they don't leave much of a hole in the jamb to fill. But the jambs will climb right up the shanks of these screws if you don't clamp the jambs to the studs.

There is some debate among designers and builders about condensation in the walls of houses that are framed with metal studs. Some folks believe that warm, moist interior air might go from a vapor to a liquid on the steel frame under certain conditions.

We decided to install a 6-mil poly vapor barrier on the inside of the frame. This became a hassle. You can't just hammer tack the stuff to the studs and joists. Instead, we tried getting it to stick by coating the studs with 3M spray adhesive. That didn't work. Both the studs and the plastic were too slippery. So we put duct-tape patches on the plastic over the studs and ran pan-head screws through them. This had to be one of the most frustrating parts of the project. My guess is that the airtight drywall approach, which uses foam gaskets behind drywall coated with vapor-barrier paint, is a better way to solve this problem.

This drill is self-supporting. An electromagnetic drill press clamps onto a workpiece. Here the author drills ½-in. holes in a ¼-in. steel hold-down and a 16-ga. diagonal strap.

My advice is to leave the hanging of the drywall to the professionals who have commercial experience. If you decide to hang your own drywall, be sure to use bugle-head screws with self-drilling points. A standard, pointed-tip drywall screw will penetrate a 20-ga. stud, but we learned that by the time the screw goes into the stud, the spinning screw has excavated a hole in the drywall that is larger than the head of the screw. We trimmed the bottoms of the walls with MDF base-

boards. We used the square-drive self-drilling trim screws to attach them, along with a continuous bead of construction adhesive.

So how do you like working with steel?—Faster assembly is one of the advantages of steel construction that turns up in most of the articles and brochures on the subject. That may be true for a simple house, but in my experience it just isn't so for a complicated custom house. The labor cost was double what we thought it was going to be.

The cost of the steel, however, was a lot less than the materials would have cost for the house to be framed of wood. All the steel, plus the fasteners required to assemble the frame, cost about $17,000 for this 1,800 sq. ft. house. The same parts in wood penciled out at about $40,000. That's a pretty impressive difference. I easily could build a simple house of steel, knowing what I know now, for less than a comparable house of wood.

For me, though, building with steel just isn't much fun. I think part of the problem is that it takes so long to fasten the frame together. When I think back on this job, there is one thing that really sticks out in my mind: There was never a time when the job took off. With wood construction there are times when changes come slowly, but then there are sweeping changes when the walls are lifted into place and the sheathing goes on. Steel construction just seems to plod along. I expect all that will change when somebody comes up with the steel-framing equivalent of the nail gun. □

Robert McCullough is a custom builder living in Seattle, Washington. Photos by Charles Miller.

Framing With
The Plumber in Mind

A few tips to help you keep your sticks and nails out of my way

by Peter Hemp

Contrary to what you might believe, most plumbers, including yours truly, do not enjoy chopping a house to pieces in order to get plumbing systems in place. However, I do claim ownership of a carborundum-tipped chainsaw for just this purpose.

When cost estimating, whether I'm on a job site or working from a set of blueprints, I first start looking for the amount of chop time that's necessary to get the rough plumbing in place. The more wood that I can leave untouched, the cheaper my labor bill is going to be.

Many things will directly influence the labor cost of installing the rough plumbing, including the location of windows, medicine cabinets, let-in bracing, beams and the HVAC ductwork. But some of the most important factors are the direction and position of the floor joists and the availability of unobstructed pathways for vents.

Even minimal plan changes might need clearance from a higher authority, so I'd like to clue you in to those areas you can improve upon with little fuss, very little cost and all by yourself.

Laying down the joists—In the past, I have severed innumerable floor joists because they interfered with toilet wastes or tub and sink drains. Having done some time on framing crews myself, I remember how easy it was to lay joists from one end of the house to the other without thinking about the location of the walls until the subfloor was down. But in relation to plumbing costs, it's here that the greatest savings in labor (mine and yours) and the biggest gain in structural integrity can be realized.

On your next set of prints, look at the location of the bathroom wall behind the toilet. I call it the "tank wall." For standard toilets, the measurement from the finished wall surface to the center of the toilet drain (or closet bend) in the floor is usually 12 in. I like to use 13 in., which gives more clearance behind the tank for future paint jobs (drawing below). I'll need a minimum 3 in. of clearance around this center point to rough in ABS plumbing (about 3½ in. for no-hub cast-iron pipe). This allows room for the pipe (typically 4½ in. in diameter), plus the added hub diameter of the closet flange (for more on toilet installation, see *FHB* #46, pp. 54-57).

If there's a floor joist in that forbidden territory, I'm afraid it's recoil-starter time. When you plan to run joists 16 in. o. c. starting from the tank wall, center your first one about 8½ in. from the finished wall. You can then add a joist on the opposite side of the drain, again placing it at least 3 in. from the center, depending on the type of pipe used. Place an additional joist

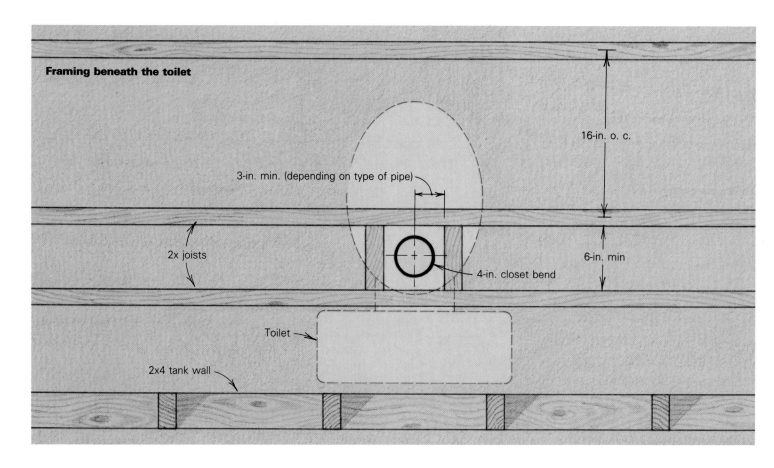

Framing beneath the toilet

16-in. o. c.

3-in. min. (depending on type of pipe)

2x joists

4-in. closet bend

6-in. min

Toilet

2x4 tank wall

Drawings: Vince Babak

wherever it's needed to satisfy the floor's structural needs.

If joists run perpendicular to the tank wall, I'll still need at least 3 in. of clearance around the center point of the closet flange. If there's a joist in the way, shift it over.

If you want to do a good deed for your customer after I've installed my drain, waste and vent system, and it has passed inspection, add some blocking around the closet drop. This blocking will stiffen the floor under the toilet, and that will help to prevent seepage from the toilet for the life of the structure. It will also save the owner lots of money later for the replacement of the subfloor and floor coverings.

Don't forget to take bathtubs into account when laying out the joists. Where joists run parallel to the length of the tub, I'd like a joist on either side of the drain hole, 6 in. from the center, with a block about 12 in. away from the shower-head (or valve) wall to support the pipe (drawing below). This distance will vary according to the type of tub and drain pipe, so check your tub before nailing up the block. When joists run perpendicular to the length of the tub, don't put a joist any closer than 12 in. to the valve wall if you can help it. If the toilet and shower head flank the same wall, you can't satisfy the requirements for both. In that case, leave a joist 8 ½ in. from the tank wall. I'll notch the top of the joist for the tub shoe (the drain fitting) if it's in the way and drill through the joist for the tub drain.

Again, you may have to add one or more joists to maintain proper joist spacing beneath the tub. But when you compare the cost of a couple extra joists to the cost of re-

placing a severed joist in not-so-perfect working conditions (how would *you* like crawling around in the dark over moist, unidentifiable objects?) you'll find the cost of the extra joists a real bargain.

Framing for tubs—While we're on the subject of tubs, most residential tubs are between 30 and 32 in. wide. When carpenters frame the valve wall, they often put a stud about 16 in. from the adjoining wall. What a shame. How many times have you sat or stood in a tub and visually lined up the strainer in the bottom of the tub with the waste and overflow plate above it, the tub spout, the tub and shower valve and finally the shower arm and head? Did it bother you to see them all out of whack? Well, this all goes back to putting that stud in the middle of the valve wall. I have to carve up the stud to anchor my valve, spout and shower arm, and it never comes off looking professional.

Instead of that one stud in the center, please install two, dividing the width of the tub into three equal spaces. That way, I can add blocking between the two studs to anchor my plumbing fixtures. The blocking also makes it easy to line up the fixtures vertically.

Troubles at the perimeter—My final gripe about floor joists has to do with those lying beneath exterior 2x4 plumbing walls. When the joists run perpendicular to the plumbing wall, I have to contend with either a 2x rim joist or a rim joist with a row or two of 2x blocking. When the joists run parallel to the wall, I'll usually find a single joist toenailed to the mudsill.

When I have to bring 2-in. drain pipes (2½-in. O. D.) down through the bottom plate of the wall, the structure suffers. For the lines to stay completely inside the wall, (drywallers love you when they don't) I have to run a 2½-in. pipe through, at best, about a 2-in. space. If there's a fitting on the drain within the height of the joist, which is often the case, I need an additional ¼ in. to ⅜ in. of clearance. What does this mean? It means I have to remove ½ to ⅞ in. from the inside of the rim joist or block (if there's an inner block, I have to tear it out first). Worse, my drill makes a 2⅝-in. hole when it's sharp. When it's dull, it travels out-of-round and adds another ⅛ in. to that. So the joist or block can wind up being just ½ in. to 1 in. thick, give or take a few hairs. That's if my vertical cutting, done with the long rough-in blade in my reciprocating saw, is perfectly plumb, which it usually isn't. And that's before I chisel out between saw cuts.

You might be wondering why I don't use smaller pipes. If the same plumber who installed new pipes in a structure had to come back later to unclog them, there wouldn't be any drain lines smaller than 2 in. in a house. This happens to be my credo. Though our local building code calls for 1¼-in. drains for lavatories, 1½-in. drains for tubs and 2-in. drains for kitchens and laundries, I prefer to use 2-in. drains for all of them. If I do use a smaller diameter drain in a 2x4 wall, say 1½ in., it still means chewing away ¼ to ½-in of wood. I could reduce the diameter of the pipe by using DWV copper (which has half the wall thickness of supply-line copper) instead of ABS plastic or no-hub iron pipe, but

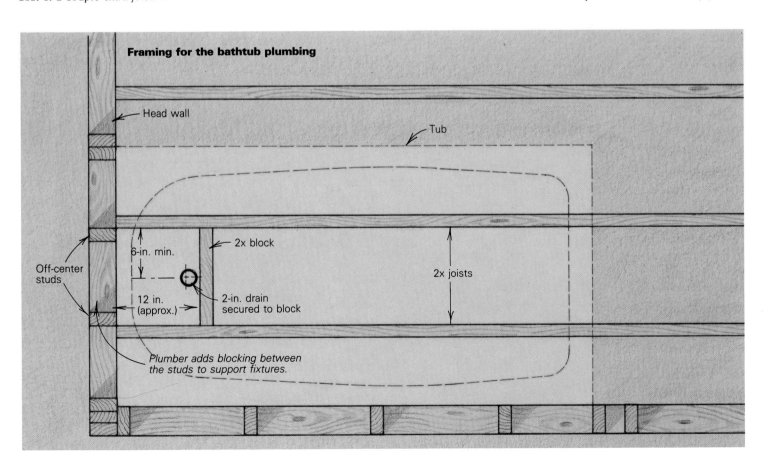

Framing for the bathtub plumbing

Head wall

Tub

Off-center studs

6-in. min.

2x block

12 in. (approx.)

2-in. drain secured to block

2x joists

Plumber adds blocking between the studs to support fixtures.

the cost of the copper pipe and fittings and of soldering the joints is considerable. Also, the torch I need for soldering could start a fire (if you don't think that plumbers are guilty of starting a few fires, think again).

So what is the remedy for all this? It's simple: furred walls. Nail 2x2s to the studs, and that extra 1⅝ in. will help me to stay clear of the rim joist or end blocks. You'll spend a few extra bucks for materials, but I'll bet in labor (yours and mine), furring the wall is cheaper than paying me to crawl and hack. And the structure of the house suffers less. If the rim joists parallel to the mudsill are doubled, which is sometimes the case, I'll still have to chew through the inner joist, but the furring will allow me to keep the outer joist intact. Of course, 2x6 walls would be even better than furring, or the designer can design the house to keep plumbing out of exterior walls in the first place.

Nails in the plumbing zone—A plumber spends most of his time boring holes. Unlike those "sparkies" who rarely have to drill any hole larger than a finger for their skinny little wires, we plumbers occasionally find ourselves boring 5-in. holes. And, the worst thing to encounter when boring big holes is a nail.

For fast drilling in wood, plumbers like to use self-feeding bits, which means we try to hang on to that powerful Milwaukee Hole Hawg, waiting to have it wrenched from our hands, pin our wrists, slap us in the face with the handle or spin us right off a ladder while the bit chews its own way through your masterwork. These drill bits can easily cost $25

or more and can be destroyed by nails the first day out of the box.

When I encounter a nail, I stop drilling immediately (as a safety feature, there is no trigger hold-down button on a Hawg; to have one would be suicide). I then have to remove the self-feed bit, replace it with a hole saw and continue drilling at a much slower rate, expending back-breaking energy. Now, if you were merely to stab your 16d sinkers a little differently, you would save me a lot of hassle and yourself some money.

Here's the program. Before you nail up a wall, consult your funny papers (some call 'em blueprints) to locate the plumbing fixtures. Next, mark the stud bays where the pipe is supposed to go, plus one bay on either side. Then, when you're nailing the bottom plates to the subfloor, start the nails in the middle of the marked bays, but don't sink them entirely. That way, if they interfere with my boring path, I can yank them out, move them over and set them myself. This also gives me the option of running the pipe through the adjacent bays if I have to. Do the same thing with the top plates; I'll sink all the nails when I'm through.

If the wall needs blocking, set the nails in the marked bays just enough to hold the blocks in place. I can bore the lower and upper plates, yank the blocks and bore them separately. Then I can slide them over my pipe (if there aren't any couplings in the way) and nail them in place myself. If the holes need to be bored near the ends of the blocks, the nails will usually split them. To avoid this, I use my cordless screw gun to predrill the blocks and screw them in place.

Backing the drywall—If you aren't asleep yet, hang on for the finale. My final suggestion deals with backing for the ceiling drywall. It used to be that a 1x6 was used for backing on top of a 2x4 wall, and a 1x8 backed a 2x6 wall. These days, two ceiling joists are often used instead, each slightly overlapping the opposite edges of the top plate (drawing below). This creates a U-channel, which can make it very tough for me to run my pipes.

If the plumbing wall is anywhere near the bottom slope of the roof, and the roof sheathing is already down, working in that location is almost impossible. The U-channel also makes it tough to tighten no-hub couplings, which are most often down in the channel. And there's more. Sometimes I try to avoid extra vents in the roof by back-venting several 2-in. vents into one 4-in. stack. The U-channels can end up getting so mauled that I have doubts as to their integrity.

On interior plumbing walls, I'd recommend using the 1x backing and avoiding center-span nailing in the plumbing bays. As I did for the plates and blocks, I'll finish the nailing for you after my pipe is through.

It may not be evident at first, but I think you'll find that my recommendations won't just save me labor, they'll save you some painful rehab activity that you would just as soon not experience. □

Peter Hemp is a plumber and writer from Albany, Calif., and author of The Straight Poop *(Ten Speed Press, P. O. Box 7123, Berkeley, Calif. 94707, 1986. $9.95, softcover; 176 pp.) a book on plumbing maintenance and repair.*

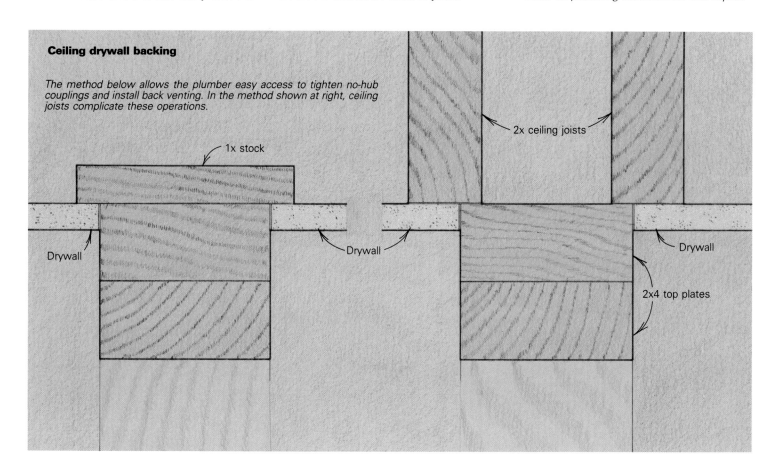

Ceiling drywall backing

The method below allows the plumber easy access to tighten no-hub couplings and install back venting. In the method shown at right, ceiling joists complicate these operations.

1x stock

2x ceiling joists

Drywall

Drywall

Drywall

2x4 top plates

Framing for Garage Doors

Think about the door before you pour

By Steve Riley

The low hum and clickety-clack rhythm of an automatic garage-door opener is a sound of our times. And yet the ingenious hardware and signal receivers that operate the doors are affixed to the most basic form in building: a beam supported by posts or, as we call them today, a header on trimmers and king studs. Making sure the framing around the door is assembled accurately and with structural integrity is the topic of this article. Building a garage and getting it ready for a door typically involves several different trades, and it takes some planning to make sure the guy who finally installs the door can do a clean job with no complaints. After all, nobody likes to chip away concrete to make room for the door tracks, and nobody wants to pay for a custom garage door because the stock one didn't quite fit in the opening.

Door types and sizes—Stock residential garage doors generally range in size from 8 ft. wide and 6 ft. 6 in. tall to 18 ft. by 8 ft. A good set of construction drawings will call out the garage-door details on the framing plans and on the elevation sheet. If the drawings lack this information, I check with the clients to find out what kind of vehicles they plan to park in their garage. For a midsized car, the typical door will have to be 8 ft. wide and 7 ft. tall. But most people want the flexibility of pulling a Suburban with a luggage rack into their garage, which requires a 9-ft. wide door that is 8 ft. tall. The extra width also reduces the chance of a driver ripping off the passenger-side mirror.

Here in the Wood River Valley in Idaho where I build houses, clients prefer sectional garage doors. They are made of four or more horizontal panels linked by hinges, with rollers that ride in tracks mounted to 2x6 side casings (drawing below right). The hardware that mounts over the door is attached to a 2x12 head casing.

Sectional doors don't mount between the jambs. Instead, they mount just inside the garage and are sealed against the weather by stops nailed to the jambs. The weather seal is one reason why folks around here prefer sectional doors. For an even better seal, the stops can include weatherstripping.

In some parts of the country, people still use one-piece, or tilt-up, garage doors (drawing middle right). These are less expensive than sectional doors and easier for builders, or homeowners, to install. One-piece doors fit between 2x side jambs that extend into the garage 2 in. beyond

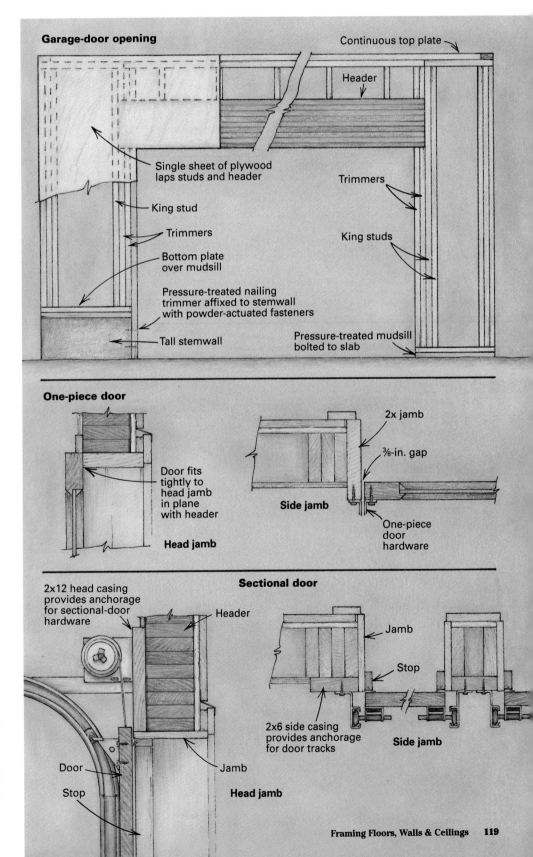

Garage-door opening

Continuous top plate

Header

Single sheet of plywood laps studs and header

Trimmers

King stud

Trimmers

King studs

Bottom plate over mudsill

Pressure-treated nailing trimmer affixed to stemwall with powder-actuated fasteners

Pressure-treated mudsill bolted to slab

Tall stemwall

One-piece door

Door fits tightly to head jamb in plane with header

2x jamb

⅜-in. gap

Side jamb

One-piece door hardware

Head jamb

Sectional door

2x12 head casing provides anchorage for sectional-door hardware

Header

Jamb

Stop

Door

2x6 side casing provides anchorage for door tracks

Side jamb

Stop

Jamb

Head jamb

Drawings: Christopher Clapp

Framing with steel

By Tom Conerly

Building designers in Santa Cruz, California, have their share of specialized design criteria. And sometimes the criteria can work to cross purposes. Take, for example, two of the conditions I had to reconcile in the design of a duplex for one of our historic residential districts.

The original duplex was destroyed in the 1989 earthquake, along with many of the neighboring homes and commercial buildings.

Fortunately, enough of the original structures remained undamaged to preserve the flavor of the old part of town, and to make sure it stays that way, any new construction on this street has to be compatible with the neighborhood. So among other guidelines, the two single-car garages required for the new duplex had to be kept separate to reduce their bulk. I put the living spaces over the garages, which fits with the row-house look along this street. But stacking a floor over a narrow garage that's mostly doorway makes for some potentially heavy loading on the garage-door framing.

As you can imagine, we have to adhere to some of the toughest seismic design standards anywhere in the world. Ordinarily, I would reinforce the corners of the garage-door wall with plywood sidewalls. But our tall, narrow building precluded that option because the garage sidewalls were too narrow to contribute much stiffness. Fortunately, our project engineer, Michael Martin, had a budget-conscious alternative for us: the moment frame (photos below)

Martin learned about steel-moment frames designing fast-food restaurants, a job that required him to reinforce openings for large windows in wood-frame buildings. A moment frame relies on a stiff connection between framing members to resist the *moment,* or forces that cause bending, in a structural member. It's difficult to make a moment connection with wood because the fasteners tend to act as pins, thereby allowing some flex. No flex is allowed in a moment connection.

As shown in the drawing at left, our moment frames are made of 3-in. by 7-in. by ⅜-in. thick steel tubing. They are bolted to the foundation with ¾-in. thick steel flanges over 1-in. dia. threaded steel rods (photo below). The rods are 27 in. long, and they have nuts and washers on their lower ends to anchor them in the concrete. The 2x wood framing members that sandwich the steel are secured by 2-in. long pieces of ½-in. threaded rod welded to the sides of the steel tubes.

We had the two frames fabricated by a local welding shop for $650 apiece, plus another $100 to transport them to the site. To ensure accuracy, our contractor, Rob Moeller, provided the shop with flange templates that gave the exact positions of the foundation bolts.

We scheduled the arrival of the frames to coincide with a crane that had to be there anyway to lift other materials onto the roof. It took an hour's worth of crane time to place the two frames and another couple of hours of carpentry to bolt the 2x stock to the steel.

I put narrow gable roofs on top of each garage to break up the façades of the duplex and to keep them from looking top-heavy. As a consequence, the upstairs walls are about 2 ft. back from the plane of the moment frame. Shear loads are transferred from the walls to the frame by way of a horizontal plywood diaphragm (for more on how wood-frame buildings react in earthquakes, see pp. 104-109).

— *Tom Conerly is a building designer based in Santa Cruz, Calif. Photos by the author.*

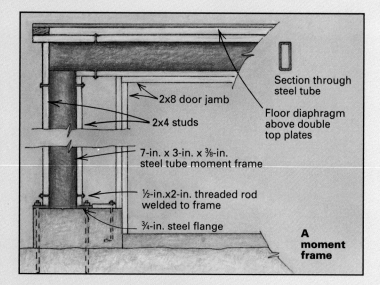

Section through steel tube

Floor diaphragm above double top plates

2x8 door jamb

2x4 studs

7-in. x 3-in. x ⅜-in. steel tube moment frame

½-in.x2-in. threaded rod welded to frame

¾-in. steel flange

A moment frame

the plane of the header. The hardware is bolted to the edge of the jamb. Many builders make the mistake of using redwood jambs for this detail. Garage-door installers prefer Douglas-fir jambs because lag bolts hold better in them. Unlike sectional garage doors, one-piece doors require a ⅜-in. gap at each side to keep them from scraping the jambs during operation.

The type of door and the hardware it uses influence the ceiling height of the garage. The minimum overhead clearance for standard sectional-door hardware is 12 in. above the bottom of the head jamb, plus another 2 in. if the installation includes an automatic door opener. Reduced-clearance tracks that fit in a 6-in. deep space can be specially ordered from most manufacturers for low-headroom situations.

Stemwall or slab—Garage walls bear on either a monolithic slab that includes a footing or on stemwalls that border a slab. In either case, make sure anchor bolts and hold-downs are accurately placed—especially in the short walls that flank the garage opening. These walls are typically 2 ft. long and require a pair of anchor bolts, each within 1 ft. of the ends. Nowadays, building codes often require metal hold-downs as well. To fit all the hardware, I make these short walls as long as possible by running them past any adjoining walls instead of abutting them.

Of the two foundations, I prefer the stemwall because it's easier to position the anchor bolts on the forms prior to the pour. If hold-downs are called for, consider the strap-type, such as Simpson's HPAHD22 (Simpson Strong-Tie Co., Inc., 1450 Doolittle Dr., San Leandro, Calif. 94557; 510-562-7775) if it satisfies the engineer. This hold-down has a hook that angles into the stemwall while the strap extends up and is anchored to the side of the studs. It's easier to deal with than hold-downs that require foundation bolts.

For a sectional door, I want an opening between stemwalls that is equal to the width of the door plus the thickness of two nailing trimmers and two jambs. For a one-piece door, the opening between the stemwalls is equal to the width of the door plus two jambs and a ¾-in. gap. I don't leave a shim space because I make sure my trimmers are plumb during framing.

King studs, trimmers and headers—Whether on a slab or on a stemwall, I prefer that the ends of my garage headers rest on two trimmers (top drawing, p. 119). With a tall stemwall (8 in. or more), I add a third pressure-treated trimmer that extends all the way to the garage floor to provide a nailing base for the bottom of the jamb and casing. I secure this trimmer to the stemwall with a couple of powder-actuated fasteners.

For a sectional door, I want at least four framing members on each side of the door opening. That way I've got backing for the drywall and for the

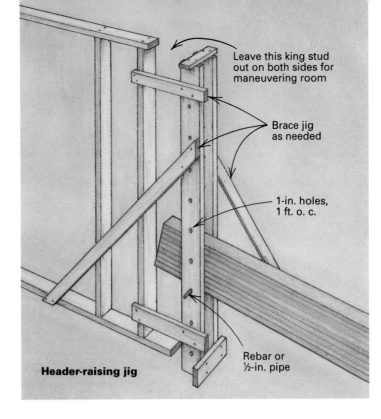

Leave this king stud out on both sides for maneuvering room

Brace jig as needed

1-in. holes, 1 ft. o. c.

Rebar or ½-in. pipe

Header-raising jig

2x6 casings that carry the tracks for the doors. The framing members can be two trimmers and two king studs or three trimmers and one king stud. When framing two, one-car sectional doors next to one another, you need a column at least 6½ in. wide between them for the door tracks (bottom drawing, p. 119).

Although the front of a garage is a continuous wall, the large opening makes it tough to erect this way. So I frame the short sections that flank the door as separate walls and connect them later with a continuous top plate. I frame these short walls with bottom plates drilled to fit over the anchor bolts holding down the mudsills. After lifting a wall into place, I make sure the outside edge of the bottom plate is flush with the face of the stemwall. This step ensures that the sheathing and the siding will be outside the plane of the concrete. At this stage, stretch a stringline from corner to corner to make sure the walls are in the same plane—stemwalls aren't always exactly where they're supposed to be.

Bottom plates aligned, I nail them to the mudsills. Then I nail the outside corners of the adjoining walls and brace the narrow walls so that they stay put as we place the header.

Custom veneer. **A sectional garage door can blend with the house when compatible siding is affixed to it. Horizontal finishes (lap siding or shingles) can make the lines between panels virtually disappear. Photo by Charles Miller.**

We can muscle a single-door header into place by hand (for example, a 9-ft., 6x14 weighs about 175 lb.). But it's almost essential to have a crane or a forklift on site to place a header large enough to span a double-wide door. Lacking a crane or a lift, we resort to a ladder-like header-raising jig that allows us to lift one end of the header at a time (drawing left). Using this jig requires the cripples and the two king studs to be left out of the walls for maneuvering room until the header is in place. I stick with either glulams or laminated veneer beams for header stock because large, sawn timbers have a tendency to twist and check. Also, I don't use built-up headers made of 2x stock and plywood because the labor required to fabricate them costs more than the extra expense for solid stock, and built-up headers are rarely as accurate as glulams or veneered beams.

Once the header is positioned, I tie it into the wall by nailing through the king stud into its ends—a typical nailing schedule is pairs of nails on 2-in. centers. Then I add my cripples and double plates to tie it all together. After plumbing and stringlining, the wall is ready for plywood. It's important to lap full sheets over the intersections of the header, the trimmers and the king studs to reinforce these areas (top drawing, p. 119). Now I can put up my 2x casings for the door hardware and call in the door man.

Because carpenters do not install garage doors regularly, it's best to have an expert install the door. Otherwise, count on watching two or three carpenters spend several hours hunched over a set of instructions while they figure them out.

A custom touch—The most economical garage doors have hollow-core panels skinned with metal or hardboard. The next step up is the same doors, but insulated. High-end stock doors are the raised-panel variety made of solid wood.

An elegant way to make a garage door match its surroundings is to skin it with the same siding that covers the house (photo below). This step adds some weight and, in my experience, works best on doors that are 9 ft. wide or less. Consult with your door installer to find out how much weight the hardware can bear.

For a substrate, I specify insulated hardboard doors with a smooth face. Then I use galvanized staples and construction adhesive to apply the veneer to the door sections. I've put siding on doors before installation, and I've sided them after installation. The second method is better because the siding adds weight to the door, making it more difficult to install. Also, the alignment of the siding stays true when affixed to a door that's in place. The installer will come back to adjust the spring tension to offset the extra weight. □

Steve Riley is a general contractor based in Ketchum, Ida.

Building Coffered Ceilings

Three framing methods

Editor's note: The following three projects hardly look the same, but they share one detail: a coffered ceiling. A coffer is characterized by sunken panels (they're usually square or octagonal) that decorate a ceiling or a vault. Though the term is generally associated with multiple panels, a proper coffer can have a single panel. The technique is thought to derive from the visual effect created in buildings where heavy ceiling beams crossed one another, and it has been used structurally and decoratively for buildings as dissimilar as neoclassical churches and the Washington, D. C., subway system.

Don Dunkley frames coffers into custom homes, typically by creating one big recessed panel. Greg Lawrence used coffering to conceal glulam beams. And Jay Thomsen used crisscrossed 1x wood strips to create the effect of sunken panels over the surface of a vaulted ceiling. —*Mark Feirer, editor of* Fine Homebuilding.

One big coffer. **Soffits girdle this bedroom to support angled coffer framing. A single, recessed ceiling coffer is the result. Photo by Scot Zimmerman.**

Single Coffer

by Don Dunkley

Among the most common ceiling details I run into when framing custom homes is the coffered ceiling. Though the term coffer encompasses a range of ceiling treatments, around here we use it to refer to a ceiling with a perimeter soffit having a sloped inner face that rises to a flat ceiling (photo left). The detail is usually found in bedrooms and dens.

The first coffers I built were usually sloped to match the roof and fastened directly to the roof framing. There was no soffit; the sloped portion of the coffer simply died into the surrounding wall. I used this method routinely for a few years—until I realized its limitations. For one thing, linking the roof to the framing of the ceiling limited the angle of the coffering to that of the roof (unless a very steep pitch was used on the main roof). Also, there was a limit to the amount of insulation that could be put into the perimeter of the coffered ceiling. Adding a soffit to the coffering solves these problems.

The soffit encircles the room and is framed so that its underside is level with the top plate. The soffit usually extends 1 ft. to 2 ft. away from the walls and offers several advantages. Framing is simplified, the pitch of the coffer can be any angle, there's plenty of room for insulation, and the flat ceiling surrounding the room can be embellished with can lights and crown molding.

The layout and the pitch of the coffer are usually found on the floor plan or the electrical plan. But before I start framing, I usually confer with the builder or the home owner to finalize the actual size of the soffit, the pitch of the coffer and the height of both the main ceiling and the soffit. Once these dimensions have been confirmed, the framing can usually be completed in a few hours.

Traditional Coffer

by Greg Lawrence

In the course of a recent remodeling project, we removed the roof from a 1,200-sq. ft. house and built a second-story addition in its place. We had to demolish the vaulted ceiling of the existing living room to make space for the new rooms above. To support those new rooms, we installed several glulam beams parallel to the exterior wall; the photo at right and the drawing below show how we coffered the ceiling to conceal the glulams.

First we wrapped each glulam on three sides with 1x Douglas fir, detailing the edges with a round-over beading bit and a router. Then we built intersecting false beams with 2x6 blocks (ripped to match the width of the glulams) and more fir. Finally, we trimmed the ceiling with crown molding. Where the molding returned off the window head casing, a striking horned cornice was created.

The resulting coffered ceiling adds a stately look to the room and nicely complements the window muntins. □

Greg Lawrence is the owner of Green River Construction in Sebastopol, Calif. Photo by the author.

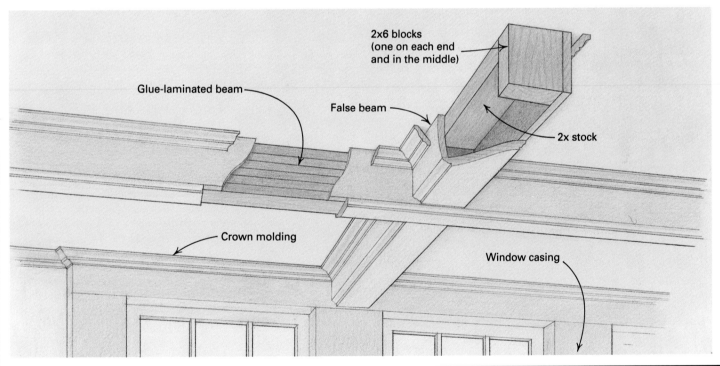

2x6 blocks
(one on each end
and in the middle)

Glue-laminated beam

False beam

2x stock

Crown molding

Window casing

It starts with the soffit—One of the big advantages of the coffering technique I've adopted is that the coffer framing can be done before the roof is constructed. That gives us plenty of room in which to work. The first step is to lay out the location of the doubled joists, sometimes called carrier joists, that form the outer edge of the soffit (drawing, p. 124). The locations are marked on the top plates of the surrounding wall, the carrier joists are oversized because they support both the coffer framing and the soffit framing—we usually use 2x10s or 2x12s, depending on the size of

the room. It's important to build this part of the framing (we call it a carrier box) straight and square. Otherwise, the rest of the coffer will be a bear to build, not to mention what the finish carpenter will say about you when he hangs the crown molding. Nail off all the carrier joists very well because green lumber, while drying, will try to go places you don't want it to visit; three nails spread the width of the boards on 16-in. centers will suffice. Of course, in order to build a good, square carrier box, the surrounding wall framing had best be on the money—a square box in an

out-of-square room will endow the soffit with a noticeable deviation in width.

To install the carrier box, start by spanning the room (usually, but not always, the shortest dimension) with doubled carrier joists. Once these have been cut and nailed in place, string a dry line across each pair and brace them straight with a temporary 2x4 "finger." Nail the finger to the carrier, push the carrier into line, then nail off the finger to the underside of the top plate. This will hold the carrier in place until the framing is complete (top right photo, p. 125). After

*A **topless hip**.* Once the soffit is in place, framing for the coffer itself is like a hip roof with the top removed. Pressure blocks are nailed between framing members on either side of the doubled carrier box; the blocks prevent the framing from twisting as it dries.

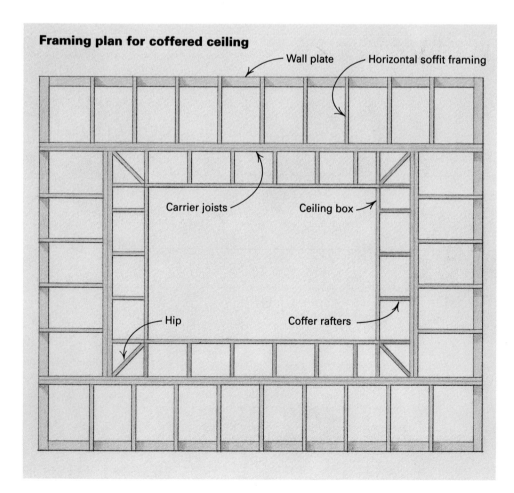

Framing plan for coffered ceiling

Wall plate

Horizontal soffit framing

Carrier joists

Ceiling box

Hip

Coffer rafters

lining the first two pairs of carrier joists, measure and hang (we use joist hangers) the second two pairs between them. These carrier joists should be lined and braced as well.

With the carrier box in place, you're ready to lay out the locations of soffit joists on the top plate. We use 2x4s 16 in. o. c. for these joists, running them perpendicular to all four pairs of carrier joists (drawing left). The soffit joists should tie into the rafters at the exterior wall plates (a code requirement in these parts), so lay the rafters out ahead of time.

As we toenail the soffit joists to the plate with 8d nails, we secure pressure blocks in every other bay (photo above). A pressure block fits snugly between the ends of the joists to prevent them from twisting as the joists dry. Nail a 1x4 to the top of the joists that are toenailed to the plate, running it the length of the wall, and secure it with a pair of 8d nails at every joist. Called a catwalk around here, the 1x4 is required by code and helps to prevent twisting at the wall end. It should be located as close as possible to the intersection of rafters and joists.

One last check for clearance—With the soffit framing in place, you're ready for the angled coffer framing—but not before one last check of the specs. If the coffer is at a steeper pitch than the roof framing to follow, now's the time to make sure that the coffer framing won't interfere with the rafters. If someone changes the pitch of the

Framing the ceiling. A ceiling box with mitered corners (photo above) forms the perimeter of the ceiling. A short hip rafter with beveled plumb cuts at top and bottom connects the corners of the ceiling box to the doubled carrier box.

Helping fingers. Pieces of scrap stock, called fingers, should be nailed between the carrier joists and the surrounding wall framing. They prevent the joists from bowing as the soffit framing is installed. Later on, the fingers will be removed.

Blocking the rafters. With the framing complete, Dunkley works his way around the ceiling to install any last pressure blocks that might be required.

roof from what's on the plans, the angle and the height of the coffer should be recalculated—a quick double-check now can avoid major problems later when the roof gets framed.

To check this, measure the run from the inside of the exterior plate (in most cases, this is where the bottom edge of the rafter will start its incline) to the inside edge of the carrier box and add this figure to the run of the coffer rafter. This gives the overall run, and by plugging this into a calculator (I use a Construction Master II) and entering the pitch of the roof, you'll end up with the height of the roof rafter's bottom edge. When 6 in. is added to account for the thickness of the ceiling framing, you'll know if the coffer will collide with the rafters. If it will, lower the pitch of the coffer.

If the ceiling height hasn't been given on the framing plans, check a section detail (if there is one). A decent set of plans usually carries all this information, but not all plans are created equal. If the plans have left this information out, you'll have to calculate the height of the coffer based on the run and pitch of the coffer rafters.

The coffer layout—The coffer layout is no mystery; think of it simply as a hip roof with the top cut off (photo facing page). At each corner there will be two common rafters and a hip rafter; the areas between corners will be filled with common rafters. After laying out a common-rafter pattern, we cut as many rafters as we'll need. Mark the locations on the carrier joists of all eight commons that form the coffer corners, then pick one corner and work your way around the box, installing the fill rafters. These are usually 16 in. o. c., but 2 ft. o. c. is fine if the coffer is small. We use either 2x4 or 2x6 stock—in general, we use what we have most of. Of course, an unusually long span might call for larger stock.

Armed with the rise and run of the coffer rafters, you can figure them for length (for more on rafter framing, see *FHB* #10, pp. 62-69). There's no need to figure in a shortening allowance, though. When the length is known, we cut one pattern and then whack out the quantity needed. If we're building more than one coffer of the same size, the second set of rafters can also be cut now.

The coffer framing—After the rafters are cut (but before installing them), we build the ceiling box at the top of the coffer, which is similar to the carrier box that forms the soffit. The difference is that the ceiling box is smaller (by the run of the coffer rafters), and the framing is not doubled up. We usually frame it on the deck from 2x6 stock, then lift it into approximate position, using temporary legs to hold it up; these legs will rest on the floor. The frame should be square; carefully cut rafters will keep it straight.

Once the ceiling box is up, install a pair of common rafters at each corner to hold the box in place. Toenail the rafters top and bottom, then install the rest of the commons, adding pressure blocks to prevent the rafters from twisting later (bottom photo, p. 125). When installing the rafters, make sure that they're not bowing the ceiling box; trim them if necessary.

When the commons are in, cut the hips to finish off the corners (top left photo, p. 125). The hips will have double cheek cuts on both ends; the cuts can be measured in place or calculated. When installing the hips, fit them in so that the drywall will follow the plane of the rafters into the center of the hip. A 6-ft. length of 1x4 makes a good straightedge to guide the hip placement. Fill in any jack rafters, if needed.

The ceiling framing is simple: Just add joists inside the ceiling box and fill in between with pressure blocks (bottom photo, p. 125). We use 2x4s laid flat to provide backing for the ceiling drywall along the length of the ceiling box. A strongback can be run down the center of the joist span to prevent the joists form sagging.

Variations—There are several variations to our coffer-framing techniques. One way to install the ceiling box is to eliminate the temporary legs and install eight common rafters at the corners of the soffit. Then lift the ceiling box up past the commons until the bottom edge is flush with the bottom of the rafters. The pressure of the commons will hold it until everything's nailed off.

Another approach is to nail the ceiling frame to the commons one board at a time, eliminating the need for help in positioning the unit. This box is supported by the hip rafters. It can withstand quite a load as long as the lower ceiling box is well braced with the ceiling-joist fingers.

Crown molding—If crown molding is desired at the top of the coffer ceiling, the ceiling joists are placed on top of the ceiling box, allowing a 5½-in. recess for the molding. To blend the bottom of the coffer rafters into the inside edge of the recess, we rip the bottom edge of the ceiling box to match the rafter slope, providing a smooth transition. If the rip reduces the width of the stock too much, cut the commons with a notch to accommodate the ceiling box so that they will blend into the inside edge (the box will be oversized by 3 in. to make up for the notch). ☐

Don Dunkley is a framing contractor in Cool, Calif. Photos by Charles Miller except where noted.

Applied Coffer

by Jay Thomsen

Usually an addition is built to reflect the design of the main house. A recent project of ours, however, showed that the opposite can also be true: Eventually, the Mrachek's house will be remodeled to reflect the addition.

As vice-president of the Handel and Haydn Society in Boston, Massachusetts, Bobbi Mrachek wanted a room in which to entertain large numbers of guests, usually for live performances of classical music. The design delivered by local architect John McConnell called for a room 40 ft. by 15 ft. 16 in., topped with a barrel vault. The look of the ceiling had to be bold, but not dark and depressing, and the surfaces had to reduce the echo effect of such a large space. Coffering would soak up the echoes. To do this without further complicating an already involved ceiling structure, we created a coffered effect with strips of 1x stock (photo facing page).

Providing the structure—As builders, we had constructed barrel vaults before, but never one so big. Given the dimensions of the new room, we knew we had a serious project to contend with. Fortunately, articles by Gerry Copeland and Lamar Henderson (see pp. 129-135) helped quite a bit.

We framed the main roof of the addition as a gable with 2x10 rafters. Flush-framed collar ties (photo top right) secured with metal gusset plates completed the basic shape of the room. Plywood gussets would form the exact curve of the vault and provide a nail base for the finish ceiling (photo bottom right).

Calculating the radius of the plywood strips was done simply by drawing a layout on the floor to exact scale using a string, a nail and a pencil. Sheets of ¾-in. CDX plywood were then laid out on top of the curve, and the radius was redrawn on top of the sheets. Three stacks of templates and several jigsaw blades later, we were ready to begin installation.

A working solution—The height of the ceiling (18 ft. 6 in.) became a factor in our planning at this point, not so much for safety reasons but for convenience: To hand each of the 1,350 pieces of the ceiling up a long ladder from below would have taken too much time. So we constructed a temporary second floor within the room to serve as a work platform—we'd need only a short stepladder to reach the highest point of the ceiling. A space 18 in. wide, running the length of the room, was left along each side of the platform to provide access for hoses, cords, passing up stock and even to dangle our legs through when working at the springline of the vault (the springline is the point of the ceiling where the curve first leaves the vertical plane of the wall).

The finish ceiling—After screwing the plywood gussets to the rafters and touching up the resulting curves with a belt sander, it was time to start nailing up the 1x6 finish ceiling. The Mracheks didn't want to see the V-groove that would char-

The shape revealed. The project began with standard gable roof framing. Flush cross ties (photo above) secured with metal plates roughed out the shape of the barrel vault. Plywood gussets (photo below) provided the final shape. They were screwed to the rafters.

acterize the seams of conventional T&G stock, so custom stock was milled from clear select pine. Every piece was prestained on both sides and both edges—twice—before installation and was sealed as soon as it went up.

We started the first piece of 1x6 at the centerline of the ceiling. Succeeding boards were then brought down either side toward the springlines. This allowed two crews of two men to work at the same time. We used nailers and 6d finish nails to secure the boards, toenailing most connections to avoid the incredible amount of time it would have taken to fill exposed nail holes with wood putty.

The coffer emerges—The main ribs of the coffering were to be layers of ½-in. pine applied in descending order of width (drawing, p. 128). Intersecting ribs would overlap each other. Our hopes were that the layering of each member

A vault of considerable size. This room addition was designed to house live performances of classical music. The lofty, coffered ceiling was created with built-up layers of 1x stock that were painstakingly screwed and nailed into place. Photo by James Shanley.

Adding up the ribs. Stepped layers of 1x stock in various widths form the ribs of the coffer. Transverse ribs are slightly wider and one layer thicker than lengthwise ribs.

Three against one. The pliable nature of pine 1x stock allowed each layer of the ribbing to follow the vault of the ceiling. Nevertheless, it took a lot of work to press each strip into place.

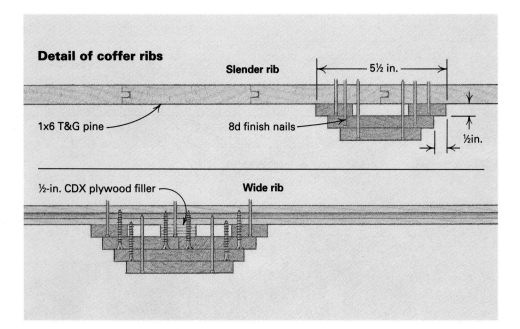

Detail of coffer ribs

Slender rib

5½ in.

1x6 T&G pine

8d finish nails

½ in.

½-in. CDX plywood filler

Wide rib

would create enough shadow lines to stand out from the background 1x6s and make the ribs seem deeper than they were (photo above). The heavier ribs that line up with the columns in the wall below are 7½ in. wide and one layer thicker than the smaller, intersecting ribs, which are 5½ in. wide. The larger ribs helped to break the ceiling into sections that were easily subdivided into smaller, coffered squares.

Fortunately, ½-in. pine conformed to the radius of the ceiling, although it took three carpenters to bend and fasten each member in place (photo left). Each layer, except the final one, was screwed in place with 1⅝-in. drywall screws; subsequent layers hid the screw heads. The final layer of each rib was secured with 8d finish nails. In some cases, especially near joints, we needed better holding power, so we used 1⅝-in. trim screws instead of nails. All pieces were installed with butt joints; we were afraid that the beveled ends of scarf joints might slide past each other as the layers were fastened into place.

Given the repetitive but very precise nature of the coffering, it was important to come up with an accurate, easy method for laying out the location of each member. Ribs were kept parallel to one another by constantly checking measurements off the end walls and by our consistent use of spacer sticks cut to the desired distance between ribs.

The layout of the ribs running perpendicular to the main ribs (parallel with the springline) was kept in line with a 10 ft. long (and very flexible) layout stick, marked with the desired rib locations. As long as the end of the layout stick was butted to the springline, and the length of the stick was snug along the ceiling curve, the layout stayed very consistent. The spacing for all ribs is approximately 2 ft. o. c.

All in all, the vaulted ceiling consumed approximately 640 man hours. If we had a similar ceiling to do again, we could probably cut about 50 man hours from the process.

All carpentry work on this job was done by carpenters Charles Desserres (lead), Brian McCune, Don Baker, Steve Harris and Mark Roberto of I. M. Hamrin Builders, Milton, Massachusetts. □

Jay Thomsen is a remodeling contractor in Milton, Mass. Photos by Charles Desserres except where noted.

Building Barrel Vaults

Two ways to ease construction with modern materials

Editor's note: Presented on the following pages are two very different approaches to building a residential barrel vault with modern materials. A vault is an arched ceiling or roof, and its precedent as an architectural form was probably set by the cave dwellings of prehistoric man. The Egyptians used bricks to build the first manmade vaults about 4,000 years ago. Engineering the vaulted form has been a recurring theme in architectural history ever since.

Masonry vaults exert significant outward thrust on their supporting walls, and for centuries the efforts to resist that thrust led to innovative construction techniques—groin vaults, pointed arches, flying buttreses. But toward the end of the 19th century, with the advent of lightweight steel frames for construction, outward thrust was becoming less of an issue in vaulted buildings. In the 20th century, the steel reinforced-concrete shell was developed. This allowed the construction of a vault that exerts no lateral thrust and can be supported just on the ends as if it were a beam. Today the engineering and construction of vaults continues to evolve in response to new building materials.

The barrel vault (also called tunnel or wagon vault) is semicircular in section and is the simplest form of vault. —Kevin Ireton

Trusses and Plywood Gussets

by Gerry Copeland

When I design a house I don't usually start with a dramatic geometric shape already in mind. Most of my designs evolve from site determinants, function, client preferences and budget. However, recently I built a house on speculation, and I wanted it to stand out from its conservative competition in suburban Spokane.

I had recently visited a small Episcopal church in Charleston, S. C. The nave of the church was dominated by a spacious great room with a voluptuous barrel-vault ceiling. This experience, along with a hankering to do some curved detail work, set me in a determined direction. Hopping on the postmodernist bandwagon, I designed a traditional gable-roofed house around a great room, with a vaulted ceiling front to back, a large Palladian window and some whimsical columns and capitals.

Design—I made the vault 16 ft. wide because that seemed to be the smallest semicircular shape that was still a functional space. Because the rear of the great room was to have a balconied loft, the curved ceiling needed to be an average usable height. I considered an 8-ft. radius to be the minimum for this.

For visual impact I wanted the space to be open from the front of the house to the back. A 16-ft. wide Palladian window at the front of the house and a 5-ft. square window at the back (selected because of cost restraints) provide dramatic lighting.

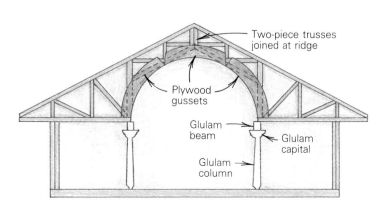

Two-piece trusses joined at ridge

Plywood gussets

Glulam beam

Glulam capital

Glulam column

The basic structure of the house seemed obvious once I laid out the central vaulted space, drew the roof pitch I wanted (6½ on 12) and added supporting columns at the bearing points. By providing a minimum of 18 in. for insulation and ventilation at the narrowest point betweeen the vaulted ceiling and the roof, the roof-truss configuration emerged. A 44-ft. long house-width truss with a bite out of the middle would have been too flimsy to transport and erect. The truss fabricator suggested dividing the truss into two at the ridge line, with two parallel-chord sections cantilevered over the vaulted space to meet in the middle (drawing left).

The design of the rest of the house fell into place quite easily, though with some compromises to the plan in order to keep the central vaulted space as a strong visual element. It was, for example, important to keep the vaulted space uncluttered by inter-

To create the barrel-vault shape, curved plywood gussets were stapled to the roof trusses. The ceiling was finished with 1x6 T&G pine nailed to the plywood gussets. The boards had to be hand-nailed because pneumatic tools wouldn't draw them up tightly against the framing.

secting wall planes. Being a builder as well as an architect, I was eager to work out the details and start framing.

Building the barrel—After all the support columns, beams and walls were in place, the trusses were set by crane onto the structure, one bundle stacked flat on top of the framing at each end of the house. Then my crew and I rolled each truss into its upright position, first one half of the pair, then the other. The cantilevered top sections straightened up nicely once they were pushed together, aligned and nailed. We stapled ½-in. plywood gussets across the adjoining webs at the ridge to tie the two trusses together. After all the trusses were erected, double-checked for alignment and nailed down, we cross-braced them according to the truss manufacturer's instructions, ran solid blocking between them over the exterior walls, and then sheathed the roof with ⅝-in. CDX plywood.

Following the wall sheathing and shingling, we cut the curved gussets that would be attached to the inside surfaces of the trusses to form the vault. After much deliberation over which material to use, we decided on ⅝-in. CDX plywood. In retrospect, a higher-quality ¾-in. plywood would have been a better choice for greater stiffness and a thicker nail base. In order to find the most cost-effective way of cutting the plywood, I spent an evening laying out the curves to scale on paper. I ended up with a cutting solution that yielded four curved gussets, 8 in. high and 8 ft. wide, per sheet.

To scribe curves on the plywood, we used an 8-ft. long wire, wrapped around a pencil on one end and around one of five nails on the other. The five nails were driven into the floor 8 in. apart and represented the centers of each arc. This allowed us to draw them quickly. Then we cut the gussets with an orbital-action jigsaw. Altogether we cut 17 sheets of plywood for a total of 66 gussets.

I'm not sure whether it was by luck or by intuition, but any three of these gussets together, point to point, made a half-circle that was exactly 16 ft. wide. So by lining up the ends of the gussets with the opening in the trusses, we could assure perfect alignment. The gussets were stapled to the trusses with 2-in. sheathing staples. When all the gussets were in place, we sighted down the 42-ft. length of the barrel and saw that the alignment was perfect. At this point in the framing, someone looked up at the exposed framing and said, "This is what the ribs of the whale must have looked like to Jonah" (photo above left).

Siding the ceiling—We used 1x6 T&G pine, prestained with two coats of semi-transparent stain, to finish the ceiling (photo above right). Keeping a straight line for 42 ft. with only a slender ⅝-in. plywood edge to nail to would not be easy, so we eliminated all the crooked pieces, as well as those pieces with loose knots. Because of the barrel-vault shape and the T&G connection, board-to-board nailing into the edge of ⅝-in. plywood seemed strong enough. But if I were to do it again, I'd use ¾-in. plywood and construction adhesive at each rib. In order to get a straight start, we chalked a line at the base of the vault, along the trusses. The first two or three courses could be nailed directly into

the bottom of each truss. We decided to use 6d finish nails driven with hammers because our pneumatic nailers wouldn't draw the boards up tight.

We quickly wished that some benefactor had donated perfectly straight and clear T&G cedar for the entire job. Every two or three rows, we sighted down the barrel and compensated for waviness by prying away from a gusset the boards that were bowed inward. All butt joints were beveled at a 45° angle. These joints looked good a year later but would have been better still had the lumber been perfectly dry.

The 16-ft. Palladian window was made for us by a local cabinetmaker and custom window fabricator. Most window manufacturers limit the size of their efforts to windows 8 ft. or less in diameter. Because of this window's size, extra thick muntins, 3-in. by 6-in., were used to withstand lateral wind load. It was built and shipped to the site in one piece, and installing it was a struggle for three of us. The double-insulated glazing was installed by window-glass fabricators after the frame was in place.

Glulam columns—Along the front half of the barrel vault, the roof trusses bear on 6-in. by 16-in. glulam beams, supported by whimsical columns cut in a classical profile (photo facing page). The columns were cut from 5⅛-in. by 12-in. standard architectural-quality glulams made up of laminated 2x6s. Because they were to be painted, minor blemishes and construction bruises could be filled, sanded and finished upon completion. We made the 8-ft. long taper cuts on each side of the columns by run-

The front half of the barrel vault is supported by a pair of 6-in. by 16-in. glulam beams, which in turn are held up by tapered glulam columns.

ning the pieces through a large bandsaw. Even with a new 1-in. wide blade we could barely cut straight enough to enable a 6-in. hand-held power plane to smooth out the irregularities. All four edges along the column's length were finished with a router and a ¾-in. beading bit, starting 6 in. down from the top and stopping 10 in. up from the bottom.

The column capitals with the tight radius cuts were done easily on the same large bandsaw using a ⅜-in. blade. These short pieces of glulam were easy to handle. We made the curving cuts so cleanly that only a minimum of sanding was necessary to finish. The edges were dressed with a ⅜-in. roundover bit. Then we attached the capitals to the columns by drilling down from the top and fastening them together with two ⅜-in. by 12-in. lag screws. The bottoms

of the capitals were notched 1 in. to sit over the tops of the columns.

My barrel-vaulted spec house definitely stood out from its conservative competition, but was on the market for an agonizing two years before it eventually sold. □

─────────

Gerry Copeland is an architect/builder in Spokane, Washington. Photos by the author except where noted.

Longitudinal Wood I-Beams

by Lamar Henderson

Drawing a simple arc on an elevation can change the entire approach to the design and construction of a project. At least that's what happened with the Gottlieb/Davis remodel. Dan Gottlieb and Peggy Davis had originally wanted a kitchen remodel and greenhouse addition to their house in Palo Alto, California. But after analyzing their programmatic needs, the existing residence and site, as well as property values in the neighborhood, we agreed that a more extensive remodel was appropriate.

The general massing of the house took on the appearance of two row houses, each reflecting a different living zone. One became the private zone: the bedroom area with a pitched roof. The other was the public zone: living, dining, kitchen, sunroom, second-story library and a bridge to the other side. We jokingly referred to the public space as the grand hall. In the search to find a dramatic roof for the grand hall, I sketched a barrel-vault ceiling and changed the course of the project.

I submitted the barrel-vault design to Dan and Peggy as one of three proposals. They were apprehensive, but at my suggestion, they decided to build a scale model, and once they were able to visualize the design, the barrel vault became the obvious choice.

We knew that the complexity of building the vault would either deter some contractors from bidding on the project or would result in highly inflated bids. So after a thorough discussion, Dan and Peggy and I decided to build the project ourselves.

Design dilemmas—Conceptually, the grand hall is a basilica with a nave and side aisles. The aisle on the south became the entry, stairs and sunroom. And the loads from the vault would flow through the wall to the foundation. The northern aisle, however, was problematic. Due to the constraints imposed by the floor plan, there were no walls that could take roof

loads directly to the foundation. A typical strategy of using a curved truss at intervals to shape the vault would require a very large and heavy beam to carry the loads to the end walls. I had to find an alternative solution.

The elevations of the two end walls dictate the shape of the roof and interior space, so it seemed logical that the structure should span the length of the hall with loads flowing down through these end walls. The problem was to find a structural member that could span 36 ft. and yet be light enough to carry by hand. A glulam beam, for example, would have been too heavy.

The metaphor that the vault was, in effect, a floor of parallel joists in the shape of a circle, suggested that some form of lightweight wood truss could span the distance. After considering the cost and weight of various engineered components, I decided on wood I-beams with solid-lumber flanges and oriented strand-board webs purchased from Structural Development, Inc. (SDI, P.O. Box 947, Los Gatos, Calif. 95031).

Attaching the I-beams to the end walls was no problem because manufacturers of metal connectors for wood construction have developed special hangers for I-beams. Developing a blocking system to tie the long I-beams together was more of a challenge. When I initially laid out the I-beams on the end wall, aesthetics ruled over good engineering. I called for the I-beams to be laid out radially so there would be a flat surface to which to attach the exterior roof

sheathing as well as the interior drywall. This created a problem because I-beams, though strong when loaded vertically, are flimsy when loaded horizontally. I planned to solve this in part through the use of blocking, cut to the curve of the arc between adjacent flanges.

When the project passed through the Palo Alto building department, the plan checker asked for a more detailed analysis of the shell structure, especially the rotation of the beams in their weak axis and the behavior of the entire roof assembly. To find out how to do this, I contacted the American Plywood Assoc. (P. O. Box 11700, Tacoma, Wash. 98411). The call turned out to be the nadir of the project. One of their engineers said the structural system should be analyzed as a curved stress-skin panel. Not only would this be mathematically complicated, but fabricating such a panel in the field to achieve the assumed structural values would be very difficult because the plywood would have to be glued to the frame under pressure.

To eliminate the need for a shell analysis and deal with the problem of having the I-beams in the weak axis, we rotated all the I-beams in their vertical axes, thus maximizing their value as structural members. Now, the problem was to create an outside curve that would allow the roof diaphragm to work, and an inside curve would create the barrel-shape ceiling.

Two-by blocking could be cut to fit the shape of the curve and attached perpendicular to the I-beams with Simpson H4 metal anchors (Simpson Strong-Tie Co., Inc., P.O. Box 1568, San Leandro, Calif. 94577). This would make the top flanges of the I-beams rigid and create a curved surface to glue and nail the plywood. However, the bottom flange and portions of the I-beam could still move laterally. We figured the bottom flange could be made rigid by using some type of metal strapping as bridging.

Because of the way the I-beams were being installed, the bottoms

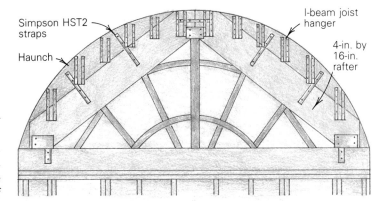

Simpson HST2 straps

Haunch

I-beam joist hanger

4-in. by 16-in. rafter

of the beams would not line up in a smooth arc and would not work as a nailing surface for the drywall ceiling. However, a quick review of the *Gypsum Construction Handbook* (United States Gypsum Co., 101 S. Wacker Dr., Chicago, Il. 60606-4385) showed a commercial system that would work for us. It employed cold-rolled steel channels and steel furring channels as a method for creating the interior vault.

After correcting the drawings, we resubmitted them to the building department for final approval. The building permit was issued and construction began.

Beaming up—The first order of business was to order the two curved glulams that would shape the end walls. We sent a drawing to a local glulam fabricator, and the bid came back at just under $6,000. Why were the glulams so expensive? Apparently, the diameter was less than the minimum allowed by the fabricating jigs, using ¾-in. stock. In order to make the glulams at such a small radius, the laminate had to be ⅜ in. thick. Material any thicker would have failed in bending.

The cost was too high. We finally solved the problem by using a pair of 4-in. by 16-in. timbers (select structural Douglas fir) as rafters, with separate pieces (called haunches) shaped to the curve of the vault and bolted on top of the rafters (drawing facing page).

The next step was to make a template of the end-wall elevation that we could use to lay out and fabricate the windows. We laid out four sheets of ½-in. plywood and struck a radius for the outside edge of the vault as well as for the head and sill of the windows. Next I drew the center 4x4 column, then the other 4x4 members dividing the window frames. I drew the width of the 4x16 rafter, locating the curve and the haunch. All metal connectors such as the straps for the haunch

and the hangers for the I-beams were also located on the template.

We cut out the plywood template, including the window openings, with a portable jigsaw. After carefully marking the pieces of the template and setting them aside, we sent the window cut-outs to the window manufacturer to use as templates for the glass.

The floor space of the second-story library was big enough to lay out and prefabricate the end walls of the vault. We cut the rafters and their haunches with a 7¼-in. worm-drive circular saw, cutting partway through from each side. We found, however, that we had to have two saws: while one was in use, the other was cooling off.

After all the pieces were cut, the end wall was assembled, checked, bolt holes were drilled for the haunch straps and hangers were located for the I-beams. First we installed the center 4x4 post, followed by the left and right 4x16 rafters, the haunch, which was bolted down using Simpson HST2 straps, and the 4x4 frames for the windows. Finally, the hangers were nailed to the rafter/haunch assembly. As the end walls were being cut and installed, the blocking and shims were being mass-produced.

With the end walls in place, it was time to install the I-beams. We started with 18-in. deep I-beams, one on each side of the vault. The length was carefully measured on each side to verify that the end walls were square and plumb. We cut the I-beams and nailed web stiffeners at the end bearing points. (Web stiffeners are vertical pieces of 2x4s cut to fit between the upper and lower chords of each truss.) The 36-ft. beam weighed approximately 150 lb.; four people could easily maneuver it into place. The next beams were 16-in. deep, weighing about 144 lb. each. The eight remaining beams were 14-in. deep and weighed about 137 lb. each. We installed the curved

blocking (16 in. o. c.) and shims as we went along in order to stabilize the I-beams (photo below). Then we crisscrossed the I-beams with metal strapping on the upper and lower flanges every 32 in. to stabilize the beams even further. At the suggestion of Kurt Anslinger from SDI, we used plumber's tape (the metal strapping that plumbers use to hang pipes) for this, which worked just fine.

According to the American Plywood Association, ½-in. plywood bent widthwise has a minimum bending radius of 6 ft., and that would work for this roof. We nailed down the sheathing with 8d ring-shank nails to the beams and to the blocking. Between the courses of blocking where the plywood edge had no support, we used H-shaped plywood clips to tie adjacent sheets together.

Before we started the interior finish work, we heeded another bit of advice from Anslinger and used duct tape to wrap the strapping where the crisscrosses touched each other. As the building moves, he had told us, the straps could rub against each other creating a bothersome noise that would be hard to fix once the ceiling was in place.

We vented the roof by drilling a series of 3-in. dia. vent holes, 18 in o. c. along the tops of all the I-beam webs. Then we installed three wind turbines on the roof and continuous soffit vents on the south side. Because there are no exterior soffits on the north side—where the addition joins the rest of the house—we added three eyebrow vents to the roof on that side. We insulated between the I-beams with 9-in. R-30 fiberglass batts.

Cold-rolled ribs and furring—Talking with a technical representative from the USG, I learned that to build a vault like this I could use 16-ga. cold-rolled steel channels for the ribs, bend them to the desired radius and then attach them to the structure. We decided to

Thirty-six-ft. long wood I-beams were supported by joist hangers on the end walls to create the barrel-vault shape. The curved blocking, run every 16 in., stabilizes the beams and provides a nailing surface for plywood sheathing.

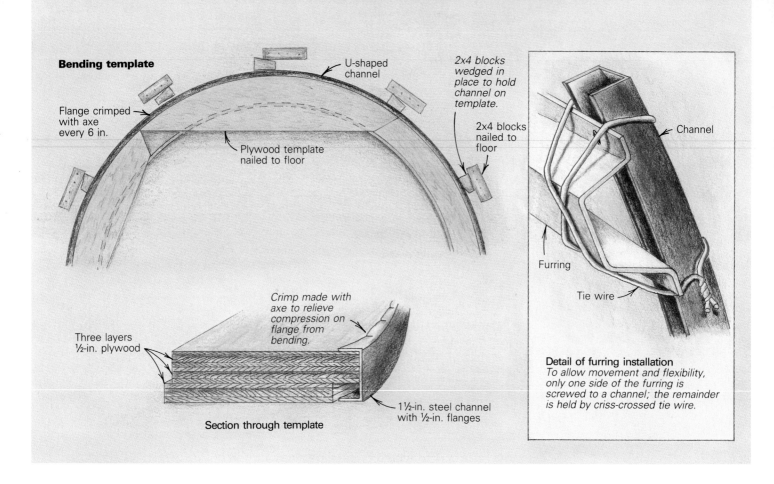

Bending template

Flange crimped with axe every 6 in.

U-shaped channel

2x4 blocks wedged in place to hold channel on template.

2x4 blocks nailed to floor

Plywood template nailed to floor

Channel

Three layers ½-in. plywood

Crimp made with axe to relieve compression on flange from bending.

1½-in. steel channel with ½-in. flanges

Section through template

Furring

Tie wire

Detail of furring installation
To allow movement and flexibility, only one side of the furring is screwed to a channel; the remainder is held by criss-crossed tie wire.

use a U-shaped channel 1½ in. wide with ½-in. flanges, installed 24 in. o. c. Then we would screw and wire metal furring every 16 in. perpendicular to the ribs. The furring, made of 25-ga. galvanized steel, is hat-shaped in section.

To find out how to bend and install this system, I called the California Drywall/Lathing Apprenticeship and Training Trust in Hayward, California (23217 Kidder St., Hayward, Calif. 94545-1632). In conjunction with their training program, they have produced a series of film strips for training apprentices on all applications of drywall, plaster and lathing. They had two films on the sequence and installation of barrel vaults.

From the first film strip I learned that bending the U-channel and maintaining its structural integrity is critical. The channel should be bent with the flanges toward the inside of the vault. Since the flanges are in compression, the material can buckle and fail while being bent, thereby losing the smooth shape of the curve. The film strip showed a technique for bending the channel using a special bending device. It also demonstrated that you should screw only one side of the hat-shaped furring to the channel. Tie wire, crisscrossed and tightened, was used to secure the remainder of the furring (detail drawing above). This holds the furring flush with the cold-rolled channels, yet allows some movement and flexibility.

The second film strip showed the installation procedure for the gypsum board. A chart (taken from the *Gypsum Construction Handbook*) listed the minimum bending radii of dry-gypsum drywall by thickness. For our vault, two ¼-in. pieces of dry material could be used, or by moistening the back paper and

thoroughly prior to application, ½-in. gypsum board could be used. The moistening was necessary to allow the back paper and gypsum (in tension) and face paper and gypsum (in compression) to stretch and compress without failure. Of course, drywall screws had to be used to attach the gypsum board to the furring.

The axe-swing dance step — Now that I knew how the vault had to go together, I had to find the bending device. I called every rental place in the San Francisco Bay Area. No luck. Then, I called lathers and drywall contractors to find out if they would bend the cold-rolled channel. Again, no luck. Instead they wanted to bid the entire job. So, how could we bend the channel?

The solution to our problem slowly emerged. We built a full-size plywood template of the arch out of ½-in. plywood, with ½-in. plywood spacers underneath, and nailed it to the floor (drawing above). The U-channel fit over it with the bottom snug against the edge of the plywood. To hold the channel in place around the template, we nailed short 2x4s every 3 ft. around the outside of the arc, leaving enough room to drive wedges between them and the template. As the channel was bent around the template, we tapped a wedge into place. The length of the arc was just under 20 ft., and since the channel came in 20-ft. lengths, only one piece was needed per rib.

Unfortunately, the channel would straighten out whenever we removed the wedges. We tried dimpling the flange every 6 in. by hitting it with a cold chisel and a hammer.

This took forever and didn't dimple the metal enough — it was hard on the template, too. We needed a faster way.

Dan happened to have an old axe in his tool box, and he decided to give it a try. By quickly hitting and denting the flange with the axe every 6 in., then turning the channel over and doing the same thing to the other side, we found that the channel would retain the shape of the arc of the template.

We had 20 ribs to bend, so we developed a rhythm for mass production. We called it the "axe-swing locomotion" dance. By rotating one's feet together from heel to toe and striking the flange with the ax, it was possible to develop a rhythmic pattern that created just the right swing motion to dent the flange uniformly. It took a great deal of hand/eye coordination and physical control on the swings to make it work, but agile dancers that we were, the job took no time at all.

The ribs were laid out at 2-ft. centers under the bottoms of the I-beams. The system was checked for roundness, plumb and parallel. The ends of each rib were securely anchored by screwing them to a metal stud run on top of the walls along both sides of the vault. At alternate I-beams, a ½-in. hole was drilled through the web, and the rib was tied to the I-beam with wire. The ribs were tied with a little slack in order to allow movement and to encourage the weight of the system to rest on the walls instead of hanging from the roof structure.

After the ribs were installed, we attached furring at 16 in. o. c. We continued with this procedure across the length of the roof, moving the scaffolding as necessary. Nine ft.

Inside the barrel vault, the drywall ceiling was screwed to a metal framework of arched ribs and straight furring hung from the bottom of the I-beams (above). Two layers of ¼-in. drywall were used to form the ceiling. The second layer was run perpendicular to the first. Residential drywall contractors shied away from the project, but a commercial crew that was between jobs happily took it on. They hung, taped and finished the job (below), including the barrel-vault ceiling, in three weeks.

Photo by Staff

from the end of each wall, at the top of the barrel vault, an electrical box was installed for ceiling fans.

Hanging the drywall—Finally, it was time to install the two layers of ¼-in. drywall on the barrel-vault ceiling. We ran the first layer parallel to the ribs (photo above) and the second layer perpendicular to the first. Soon after we started applying the drywall, our two laborers left the job to return to school. Because this was one of the least-pleasant tasks of the entire project, we decided it was time to call in a professional.

Most residential drywall installers weren't interested in working on the project because of the metal furring system, the ceiling height and the overall complexity of the project. However, we were lucky to find a commercial installer who needed to fill a gap in his work schedule to keep his crew employed. They hung, taped and textured the house in three weeks.

After they finished, we still had to cover the nuts and bolts where the straps held the haunch to the rafters. We found a local firm (San Francisco Victoriana, 2245 Palou Ave., San Francisco, Calif. 94124) that makes decorative trim plaster castings for Victorian houses and ordered 12 plaster keystones. They were hollow on the back and fit nicely over the exposed bolts (photo right).

It was hard to believe that the simple arc sketched on an exterior elevation 18 months before was now a three-dimensional reality.□

Lamar Henderson is an architect in Palo Alto, Calif. Photos by the author except where noted.

All Trussed Up

Two structural systems that blend aesthetics and engineering

Steel-And-Wood Trusses

by Eliot Goldstein

We were at a critical point in our design of the Siegel house. All the rooms seemed to be in the right places, but we still hadn't been able to resolve the design of the entry hall. The essential components of the space—a grand staircase, high ceilings and lots of daylight—had been identified, but there was still some question about how to put it all together. We needed a dramatic element to bring the space to life. And we also had to address the structural forces at work on the entry-hall roof.

The two perpendicular wings of the house would meet at the entry hall, its steep gabled roof running from the front door to the rotunda-like family room. The outward thrust of the gable rafters (they would span more than 16 ft.), would have to be resisted somehow. Though the structure throughout the rest of the house would be hidden from view, I thought exposed trusses and a cathedral ceiling seemed to be the answer here. I decided on a truss system to resist the outward thrust of the gable rafters, and my codesigner on this house (and wife), Risa Perlmutter, agreed that such a system would keep the entry hall bright and open. Trusses would also be relatively economical and could be assembled on site by the framing carpenters. Finally, a truss system would present the opportunity to create the dramatic element we'd been seeking (photo left).

Design and engineering—My initial sketches of the truss showed all of its members in wood. Our structural engineer, Nandor Szayer, suggested that we consider instead using wood for compression members and steel for those in tension. After my clients endorsed the idea, he began to work out dimensions and connections (drawing facing page). We wanted to keep the truss members as delicate as possible, so it was important that they carry only axial (or longitudinal) loads. Unfortunately,

The entry hall. **The author wanted to create a sense of drama in the entry hall and still needed to resolve the structural elements of its roof. Addressing both issues resulted in separate rafter and truss systems, with the cedar ceiling seeming to float unsupported.**

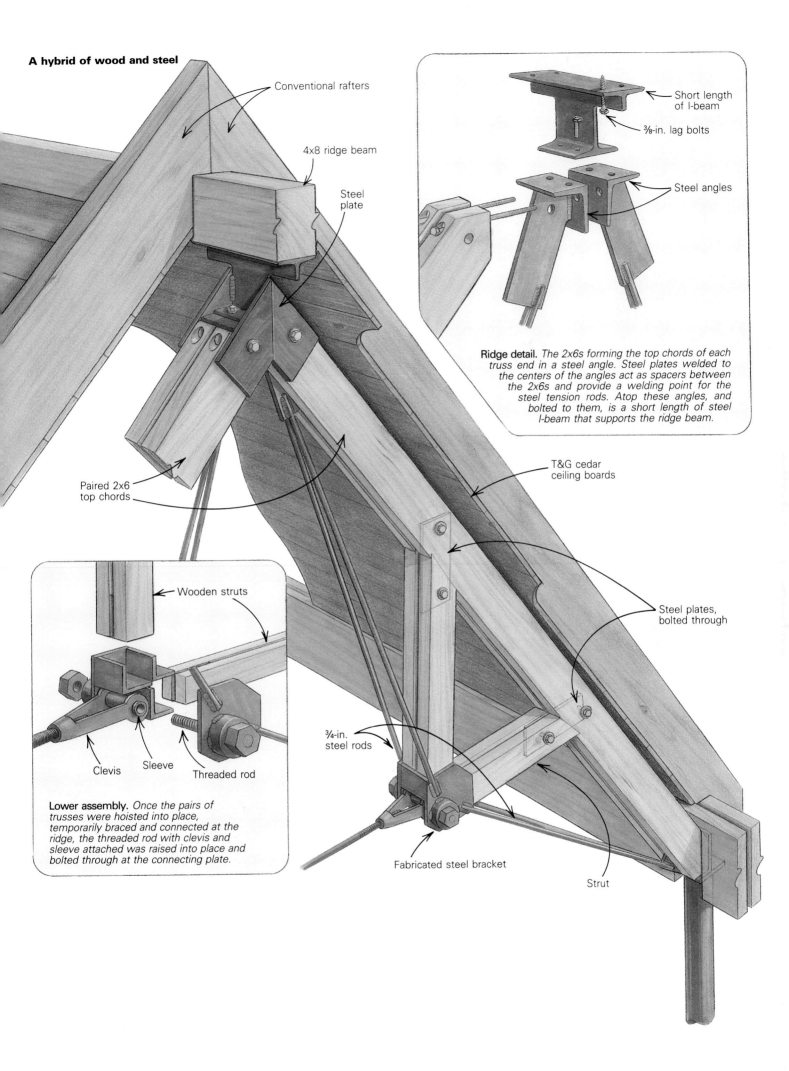

A hybrid of wood and steel

Conventional rafters

4x8 ridge beam

Steel plate

Short length of I-beam

⅜-in. lag bolts

Steel angles

Ridge detail. *The 2x6s forming the top chords of each truss end in a steel angle. Steel plates welded to the centers of the angles act as spacers between the 2x6s and provide a welding point for the steel tension rods. Atop these angles, and bolted to them, is a short length of steel I-beam that supports the ridge beam.*

Paired 2x6 top chords

T&G cedar ceiling boards

Steel plates, bolted through

Wooden struts

Clevis

Sleeve

Threaded rod

¾-in. steel rods

Lower assembly. *Once the pairs of trusses were hoisted into place, temporarily braced and connected at the ridge, the threaded rod with clevis and sleeve attached was raised into place and bolted through at the connecting plate.*

Fabricated steel bracket

Strut

top chords of roof trusses are usually uniformly loaded along their entire lengths, inducing combined loading (compression *and* bending), which requires greater member depths.

To avoid combined loading, we chose to have the truss support only the ridge beam. The rafters, in turn, would bear on that beam, and on another parallel beam at their lower ends, but not on the trusses themselves. We decided to leave a substantial gap between the bottom edges of the rafters and the top edges of the trusses so the cedar ceiling boards could be installed between them later on.

It's easy to overlook the fact that the stresses in a truss vary constantly, and that under certain circumstances (in heavy winds or under seriously unbalanced snow loads, for ex-

ample), there may even be stress reversals (from tension to compression or vice versa). Consequently, Nandor had to analyze the individual trusses for various loading conditions.

We also wanted to ensure adequate lateral bracing parallel to the plane of the truss system. We sheathed the south gable wall with plywood to create a shear wall, and tied the lower ends of the rafters above the entry hall into the plywood-sheathed attic floor (of the house's two converging wings) to prevent distortion of the gable.

Making connections—We resolved to use *concentric* joinery as much as possible, as *eccentric* connections (those in which the force lines do not converge to a point) cause twist-

ing of the joints, and ultimately, bending in the truss members. Also, it was as important to us aesthetically that our connection details be spare as it was to keep the truss members delicate. Attaining elegant, concentric joints proved a challenge.

Nandor proposed that the trusses be assembled as a pair of trussed-rafters (each of which, though only half of the completed truss, functions as a truss itself), and that they be tied together with a horizontal rod (the bottom chord), once in place. Five members would converge at each of the joints held by this rod. The conventional way of connecting many converging members is to extend a plate or other steel fabrication out from the point of convergence, in the plane of the truss, so that the

Fabricating Box-Beam Trusses
by Edwin Oribin

The first 25 years of my professional life were spent designing buildings in the tropical region of coastal northern Australia. Then, several years ago, with our children all grown up and on their own, my wife and I moved 1,500 miles south—away from the equator—to a comparatively colder and dryer climate.

Taking these climatic factors into consideration, I knew I'd have to incorporate insulation and passive-solar principles in the design of our new house. Also, as I'd be working on the house mostly by myself, I wanted the structure to be reasonably manageable. Of particular concern to me was the design for the roof system. I decided on plywood box-beam trusses because they'd be lightweight, strong and fairly economical. This type of truss would also permit me a more sculptural design.

The main consideration in designing the clamping jig for my roof trusses was the short glue-setting time summer allows. According to the manufacturer, at 86° glue should take 45 minutes to set. In fact, when I tested it, it turned out to be more like 35 minutes. I was pretty certain that even with the simplest, quickest jig I could devise, we wouldn't be able to glue, assemble and clamp down the entire truss half in less than 35 minutes. I decided to do it in two stages instead.

Building the jig was a straightforward affair. I marked out the shape of the truss half on a temporary particleboard table, then edge-glued and screwed short 2x3 blocks around the perimeter and opposite each upright to define the form (drawing above). Next, I drilled holes adjacent to these blocks and fastened a ⅜-in. dia. threaded rod above and below the table. I left the rod projecting far enough above the blocks for nut, washer and another short piece of 2x3 (with holes drilled at each end) to act as the top member of the clamp.

Preparation—Cutting to length the 1x3 top and bottom chords of the trusses went quick-

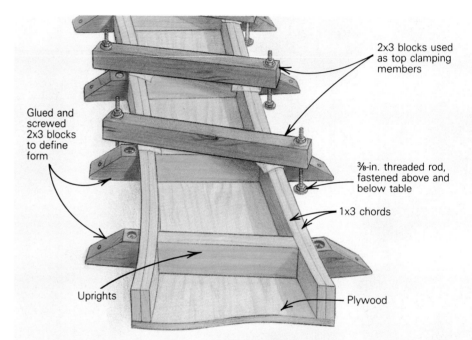

A site-built jig. *Using two sheets of particleboard end to end as a jig table, the author edge-glued and screwed small blocks of 2x3 to the table opposite each upright member to outline the form of the truss halves. Longer blocks of 2x3 span the width of the truss half and are bolted down through the table, forming the top member of each clamp. Each truss half required a two-stage glue-up because of the glue's extremely quick setting time in the summer heat.*

ly. Then came the big job of cutting all the uprights. I took angles and lengths directly off of the jig, cut every piece on the radial-arm saw and numbered and stacked the pieces of each truss in a separate pile. I used a bandsaw to cut the irregularly shaped blocks that sit on the posts (drawing facing page). Our house needed 12 trusses—all told, I cut 408 pieces of timber and 96 sections of plywood (two pieces were required for each face of each half). I cut the plywood slightly undersize so that each section would drop smoothly into place between the blocks of the jig.

Having decided to assemble the truss halves in two stages, I still wanted to do a time trial. Working with my wife, we did a simulated dry run, complete with alarm clock. The simulated "glue-up" took us about 30 minutes. Feel-

ing confident that we could duplicate the effort, we prepared to begin the real glue-up.

The urea formaldehyde glue powder we used had first to be mixed with water. It's hard to get all of the lumps thoroughly dissolved, but a rotary egg-beater did the trick. Fortunately, setting time doesn't begin until the hardener is added, but at that point there's only time for a quick final stir—then you've got to move fast.

Assembly—The first stage in our assembly procedure was to set the bottom sheets of plywood in the jig, then the pairs of 1x3s for the top and bottom chords, followed by all the uprights. All were glued as they went in, then clamped down (drawing above). I spread wax paper between the truss uprights and the 2x3 clamping blocks to keep the truss from be-actual structural connections don't interfere with one another. This approach, while structurally sound, didn't strike us as particularly interesting or expressive.

We chose an alternate solution, one which involved layering the connection. Each member was assigned a separate layer within the thickness of the truss, thus allowing all members to converge at, or very near, a single point (as seen in elevation). Each joint consists of two pairs of wood struts (one pair horizontal, one pair vertical) housed in a fabricated steel bracket, with two pairs of diagonal rods welded to the outside faces of the bracket. A Y-shaped clevis is bolted to the bracket, and a horizontal tie rod threads into the other end of the clevis. The two ends of the tie rod are threaded in opposite directions to allow tensioning (drawing, p. 137).

Assembly and installation—To ensure consistent results, the fabricator welded most of the steel components together in his shop and primed them, which would prevent corrosion while the members were exposed to the weather. On site, the carpenters simply cut and drilled the 2x lumber, then bolted it to the steel.

The steel columns on which the trusses sit were already in place. The trussed-rafters were lifted up and bolted to the cap plates of the columns; temporary bracing was installed to keep the trusses from falling like dominoes. After the tie rods had been adjusted to the proper length, the 4x8 wood ridge beam was installed. Conventional rafters were then birds'-mouthed onto the ridge and eave beams, then sheathed with plywood. A continuous ridge vent was installed.

The finished ceiling boards were not installed until the house was tight to the weather. The carpenters had left a larger gap than we'd planned between the undersides of the rafters and the tops of the trusses, but we actually like it better. The visual separation of the primary structure (the trusses) from the secondary structure (the rafters) parallels their functional distinctions. □

Eliot Goldstein is an alpine design consultant and a partner in the firm of James Goldstein & Partners, Architects, in Millburn, N. J.

The house completed. The author wanted a simple, economical, roof support system that had character. The sculptural flair of opposing curves is evident here.

coming glued to the jig. When the glue had set we undid the clamps, glued and fitted the top sections of plywood and resecured the clamps.

Once out of the jig, each truss was finished around the edges with a router set to cut the plywood ¼ in. inside the line of the timber. That gave the trusses a nice rabbeted edge. I applied one coat of stain/sealer, fitted the purlin support blocks and stacked the trusses on edge with spacer blocks between.

I would hate to pay for trusses like this made in a factory, but the cost of the materials was very reasonable, and I had the time. In my situation, it was a very economical way of spanning 20 feet. □

Edwin Oribin is an architect in Stanthorpe, Queensland, Australia.

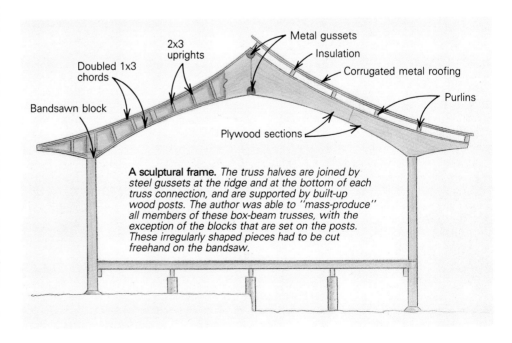

Doubled 1x3 chords

2x3 uprights

Metal gussets

Insulation

Corrugated metal roofing

Purlins

Bandsawn block

Plywood sections

A sculptural frame. The truss halves are joined by steel gussets at the ridge and at the bottom of each truss connection, and are supported by built-up wood posts. The author was able to "mass-produce" all members of these box-beam trusses, with the exception of the blocks that are set on the posts. These irregularly shaped pieces had to be cut freehand on the bandsaw.

Toenailing

by Albert Treadwell

One of the most frequently ignored devices in all of carpentry is the simple nail. Everyone takes it for granted. Everyone thinks that everybody else instinctively knows what nail to use where, and how to drive it properly. Even on the few occasions when the experts do decide to write about nailing, they rarely say much about toenailing. For example, a letter in *Fine Homebuilding* magazine from engineer/builder Don McNeice treated load engineering and design considerations for fastening joists to beams and headers, but he did not address the role of toenails, a vital element of any nail-fastened structure.

McNeice correctly points out that spikes into the end grain of a framing member have no withdrawal resistance, and it's pretty clear that a lag screw won't grab in end grain either. Friction can be an important part of a joint's strength, above and beyond the sum of the shear strength of the individual nails mandated by the design load. Properly installed toenails will pinch a joint tight, and their shear strength will help keep it tight (drawing, right).

The importance of accurate cutting to length and a good square end cannot be overemphasized, and this alone is enough to warrant using a radial-arm saw on site. All too often in today's (and yesterday's) construction, you find crooked or short rafters or joists with contact reduced to a line or a single point. Some even stand open completely, with nail shanks visible. Nails in such a joint are subject to stresses considerably more complex than simple shear, and the members are free to twist and move.

You could drill to lessen the chance of splitting, but in the case of NcNeice's face-nailed spikes this would also decrease the already minimal compression-friction grab the joist end offers the nail. On the other hand, you can drill toenails for the full diameter of the nail, because the head is pulling the joint together. When you're toenailing, you don't need to drill into the second piece, because the face grain usually takes nails well. It is possible by pre-drilling the board to

Begin by placing stud ¼ in. off layout mark.

Framing members are pinched together with final blow.

At least half a nail's length

Face-nailing

Opposing toenails have great shear strength and also resist withdrawal.

Toenailing

The strength of the joint in shear relies solely on the strength of the nails or spikes. There is little resistance to withdrawal.

Frances Boynton

place toenails anywhere you like without splitting the wood. You can drive as many as six toenails in a 2x4, eight in a 2x6, ten in a 2x8, and so forth.

Now an 8d or 10d toenail doesn't offer the shear strength of a face-nailed 20d or 30d spike, but ten 8d toenails and one or two spikes in a 2x8 will make a stronger and more durable joint than peppering the same joist with four or five spikes driven into the end grain. The strength of such a joint is simply the sum of the shear strengths of the spikes. It relies on strapping or the nailed-down floor for its integrity. The toenailed joint adds friction, developed in the joint by the pulling action of the opposing toenails, itself a result of horizontal shear in the small fasteners. Also, with toenails on all sides, force from any direction will make the joint tighter and stronger.

How to do it—Many experienced framers don't use toenailing either because they haven't thoroughly understood its practical value or because they never learned efficient technique. Poorly done, toenailing will result in split wood, misaligned work, and joints standing open, to say nothing of smashed fingers and hammers thrown into the woods.

Toenailing is not difficult and can even assist in the final alignment and positioning of the piece. To begin, place the stud or joist end in position, but off its layout line by ⅛ in. to ¼ in. Start a toenail on one side while bracing the other with your hand or foot. You may start the nail perpendicular and swing it to the correct angle, but it must be placed so that more than half of its length will enter the second piece. The nail must enter the piece to which you are joining

the stud or joist before the stud or joist is pushed into its final position. The final hammer blows will move the piece onto the line, and this is what creates the pinching action that forces the pieces tightly together. The distance you start from the layout line is critical. With some practice, you'll be able to determine this based on the size of both members and the length of the nails. Too little distance and you will either generate no pinching action or overshoot the mark, causing the piece to stand off the joint when you drive it back. Too much distance and you won't make it to the mark, and you may even split the piece by driving the head of the nail too far.

The sequence of subsequent nails is determined by the alignment corrections that are necessary. Remember that when the piece is held by one nail, it can pivot, but it won't move if force is aimed directly at that nail. So if the piece is a little crooked, you have to position the next toenail to bring it around. If you are right on, aim the next nail almost directly at the first from the other side. Continue with the remaining nails in like fashion.

Another subtlety concerns the final hammer blow to each nail; if it strikes both the nail and the wood, the piece will move. This is useful when correction is necessary. If the final blow of the hammer strikes only the nail, the piece will not move. A hammer without a crowned face and beveled edge (like my old Stanley #41) is useful here. So is a heavy nail-set (I use a ¼-in. mechanic's punch).

In a sensitive spot, it pays to drill a pilot hole. Some would say this is a waste of time, but a split destroys any strength the joint might have had. You either have to replace the split piece or leave a structurally inadequate member in place. I like to use a cordless drill, because being tethered to an outlet can be irritating and dangerous. The small holes require only a little power, so a single charge should last all day. For smaller jobs a hand drill or push drill is good. □

Contractor Albert Treadwell lives in Sandy Hook, Conn.

INDEX

The articles in this book originally appeared in *Fine Homebuilding* magazine. The date of first publication, issue number and page numbers for each article are given at right.